# ELECTRONIC LABORATORY TECHNIQUES

# ELECTRONIC LABORATORY TECHNIQUES

BY

L. W. PRICE
M.A., C.ENG., M.I.E.R.E.

*Technical Officer, Department of Biochemistry*
*University of Cambridge*

J. & A. CHURCHILL LTD.
104 Gloucester Place, London
1969

First published 1969

Standard Book Number 7000 1388 1

ALL RIGHTS RESERVED

This book may not be reproduced by any
means, in whole or in part, without
permission. Application with regard
to copyright should be addressed to
the Publishers.

Printed in Great Britain

# Contents

| | | |
|---|---|---|
| *Preface* | | vii |
| *Acknowledgements* | | viii |
| *Chapter One* | Basic Electronic Devices and Circuits | 1 |
| *Chapter Two* | Electrochemical Instruments | 27 |
| *Chapter Three* | Spectrophotometric Instruments | 81 |
| *Chapter Four* | Spectrometry | 139 |
| *Chapter Five* | Electronics and Nuclear Measurements | 175 |
| *Chapter Six* | Electronics in Analytical Separation Techniques | 221 |
| *Index* | | 255 |

# Preface

The purpose of this book is to provide an outline of the electronic techniques used in modern laboratory instrumentation. There are many books on laboratory techniques and methods of analysis, but few of these books give any details of the electronic circuits used. Modern instruments used for scientific research or analysis are becoming increasingly complex electronically and in order that the research worker can fully understand the limitations and capabilities of his instruments a working knowledge of the principles involved is vital.

Many of the operating difficulties encountered on laboratory instruments are due to simple faults which can readily be dealt with by a laboratory technician or are due to a lack of understanding of the instrument by the operator. It is hoped that the practical information in this book will therefore be of value to technicians, students and research workers.

Throughout the book, various types of instruments commercially available have been used to illustrate the basic electronic principles involved in the operation of laboratory apparatus. It should be made clear that the instrument chosen to illustrate a particular method of operation may not be the only model available and as it was not intended to carry out a survey of available laboratory instruments it has not been possible to mention all manufacturers or instrument models.

Both valve and semiconductor circuits are described, as in addition to older apparatus in regular use, a considerable number of instruments available today still employ thermionic valves. References are also made to many techniques and instrument modifications described by research workers in the literature.

The illustrations in the book have all been specially drawn by the author, although they are of course based on information supplied by manufacturers. The publication of circuit information in this book does not imply freedom from any patents held by the relevant instrument manufacturer.

Chapters Two, Three, Five and Six are based on articles originally published in *Laboratory Practice.* Some data from the series 'Electronic Techniques in Liquid Scintillation Counting', published in *World Medical Electronics,* is included in Chapter Five.

# Acknowledgements

The author wishes to thank the following instrument manufacturers or their agents for assistance with technical and circuit information during the preparation of this book. I would also like to express my appreciation to Professor Malcolm Dixon of the Department of Biochemistry at the University of Cambridge for advice given on the preparation of the articles on which this book is based. Finally, thanks to my wife for her patience in typing the draft and final manuscript of the book.

American Instrument Co. Inc.
Analogue Devices Ltd.
Applied Physics Corp.
Ataka & Co. (U.K.) Ltd.
Baird Atomic
Beckman Instruments Ltd.
Beckman Instruments International S.A.
Bellingham & Stanley Ltd.
Bendix Electronics Ltd.
Continental Distributors Ltd.
Degenhardt & Co. Ltd.
Durrum Instrument Co.
Ekco Electronics
Electronic Instruments Ltd.
E.M.I. Electronics Ltd.
F. & M. Scientifica
G.E.C.–A.E.I. Electronics Ltd.
Griffin & George Ltd. (Centrifuge Division)
Hewlett Packard Ltd.
Hilger & Watts Ltd.
V. A. Howe & Co. Ltd.
Infotronics Corporation
Instrumentation Laboratories Inc.
Intertechnique Ltd.
Isotope Developments Ltd.
Leeds & Northrup Ltd.
L.K.B. Instruments Ltd.
Measuring & Scientific Eqpt. Ltd.
M.E.L. Equipment Co.

Metrimpex Instruments
Microtek Instruments Inc.
P. K. Morgan Ltd.
Mullard Ltd.
Nuclear Enterprises (G.B.) Ltd.
Nuclear Chicago Europa
Optica (U.K.) Ltd.
Packard Instruments Ltd.
Panax Equipment Ltd.
Perkin Elmer Ltd.
Philips Scientific Instruments
Phoenix Precision Instrument Co.
Picker Nuclear Corp.
W. G. Pye Ltd.
Radiometer A.S.
Radyne Ltd.
E. H. Sargent & Co. Ltd.
Shandon Scientific Co. Ltd.
Shimadzu Seisakusho Ltd.
Southern Analytical Ltd.
Tracerlab Ltd.
Tracor Inc.
Techmation Ltd.
Techtron Ltd.
Tektronix (U.K.) Ltd.
Unicam Instruments Ltd.
Varian Aerograph Ltd.
Varian Associates Ltd.
Yellow Springs Instrument Co.
Carl Zeiss Ltd.

# 1 Basic Electronic Devices and Circuits

This first chapter is intended as an introduction for readers not normally familiar with electronics, and circuits are described that are of particular importance in laboratory instruments. The descriptions given are brief and the use of mathematics is avoided, as the intention is to outline the operation of these circuits, in order that their function in the laboratory instruments described in later chapters may be properly understood. It is of course impossible in one chapter, to describe all aspects of valve and transistor circuit operation, and for readers who wish a more detailed account, a list of books for further reading is included at the end of the chapter.

## THERMIONIC VALVES

**Valve operation**

The thermionic valve or electron tube consists basically of an evacuated glass envelope containing a heated cathode and an anode plate. Electrons emitted from the cathode are attracted to the anode when it is at a positive potential with respect the cathode. If the anode is made negative, then the emitted electrons are repelled and no current flows through the tube. It can then be seen that this two-electrode device, or diode, may be used to rectify an alternating potential applied to the anode, so that current will only flow in one direction in the external circuit.

The flow of electrons between the cathode and anode of a thermionic valve may be controlled by means of a wire mesh grid placed between them. This type of valve is known as a triode. A small alternating voltage on the grid gives rise to a corresponding change in the anode current of the valve, which, in flowing through a resistance in the anode circuit, produces a voltage change much larger than the input voltage on the grid. The ratio $\dfrac{\text{change in anode voltage } V_a}{\text{change in grid voltage } V_g}$ is known as the amplification factor u.

Several valves may be connected together, as shown in figure 1.1., to form an a.c. amplifier using resistance capacitance coupling. The valve anode current consists of a d.c. component due to the supply voltage, and an a.c. component due to the input signal. The a.c. component is fed to the following stage by the coupling capacitor $C_c$. The resistor $R_g$ allows any electrons which may be collected on the grid of $V_2$ to leak away. The valve operating point or bias is determined by the d.c. current flowing through the cathode resistance $R_k$. Capacitance $C_k$ is a bypass for the a.c. component of the valve current which, if allowed to pass through the cathode resistance, would cause degeneration or loss of gain.

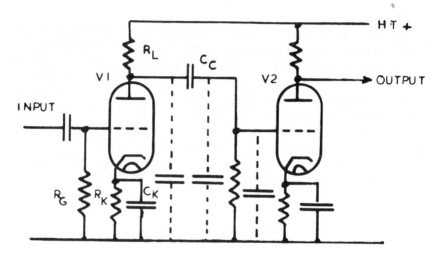

*Fig. 1.1.* Valve resistance-capacitance coupled a.c. amplifier.

The capacitances shown dotted in figure 1.1. represent the valve inter-electrode capacitances which have a negligible effect on the amplification or gain at low operating frequencies, but above about 3,000 Hz have a significant effect. In order to overcome the reduction of gain due to the inter-electrode capacitances at the higher frequencies, an electrostatic screen may be placed between the grid and anode. This type of valve is known as a tetrode, but has a 'kink' in its anode characteristic curve, due to secondary electrons emitted from the anode travelling to the positive screen and reducing the net anode current. This disadvantage is overcome in the pentode valve, which has an additional suppressor grid placed between the screen and the anode. The suppressor grid, which is negative with respect to the anode, retards the secondary electrons produced and diverts them back to the anode.

**Gas filled valves**

The types of valve so far described are known as hard valves because of the evacuated envelope. Other types of valve are however filled with gas and known as soft valves, examples being high voltage rectifier diodes filled with mercury vapour or xenon, and voltage reference or stabiliser tubes filled with neon. This latter tube does not have a heated cathode to produce the electrons and is known as a cold cathode tube.

Another soft valve is the gas filled triode or thyratron, which is controlled by the voltage applied to its grid. When this voltage is less than a certain critical negative value, current flows in the tube and the gas is ionised. The grid which is still negative, attracts positive ions which neutralize its negative field, so that once ionisation occurs the grid can no longer control the current flow. This is limited only by the applied voltage and the external load. Once the thyratron has 'fired', the current can only be stopped by opening the anode circuit or by making the anode negative, both these methods being used to bring about de-ionisation. The thyratron is used extensively for controlling electric motors and power loads, some control circuits being described in chapters 2 and 6.

## COLD CATHODE VALVES

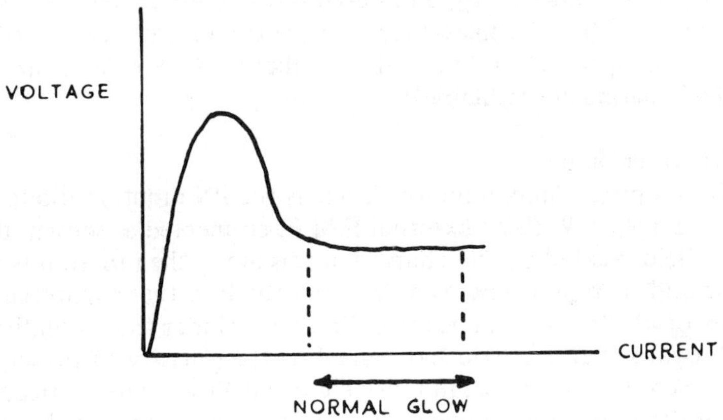

*Fig. 1.2.* Cold cathode stabiliser tube characteristic.

### Stabiliser or reference tubes

The cold cathode stabiliser reference tube has an anode and a cathode and as voltage applied to the tube rises, a point is reached when the gas filling breaks down and the current through the tube increases due to the onset of ionisation, a region occuring when the gas glows normally and is fully ionised. In this region it will be seen from figure 1.2. that the voltage across the tube remains constant.

### Corona stabilisers

Another type of constant voltage tube is the low current corona stabiliser. The tube is filled with gas at a pressure of a few millimeters of mercury and is operated in the range 400 to 1,000 volts with current up to 100 micro-amperes. It finds application in low power E.H.T. supplies of nuclear counters.

**Trigger tubes**

The cold cathode trigger valve has a third trigger electrode which is biased so that the tube just does not 'fire'. When a small pulse is applied to the trigger electrode a discharge is initiated in the tube, and since the current builds up very quickly this device may be used in high speed triggering circuits.

## SEMICONDUCTORS

Semiconductor devices are built up from P and N type materials. N type material is obtained when a pentavalent substance (e.g. arsenic) is introduced into germanium or silicon giving rise to a number of free negative charges. P type material is obtained if a trivalent substance (e.g. indium) is introduced. There is then a deficiency of electrons to complete the bond, and a number of holes or equivalent positive charges are available.

**The rectifier diode**

The simplest semiconductor device is the PN junction diode shown in figure 1.3(a). With an external E.M.F. connected as shown, the internal field created by the charged atoms along the junction is reduced and there is a flow of holes from the P to the N material with a flow of electrons in the reverse direction. Under these conditions the forward characteristic is obtained. If the polarity of the supply is reversed only a very small reverse current flows. The semiconductor diode is thus the equivalent of the two electrode valve and can be used as a rectifier.

**Zener diodes**

With the PN junction reverse biased the small current is due to minority carriers, i.e. the positive holes in the N type material and the electrons in P type material. If the reverse voltage is increased, the internal electric field is increased and a point is reached when the co-valent bonds between the atoms are torn apart and a large number of holes and electrons are set free. In this zener region as it is called, the voltage is virtually independent of the current, as can be seen from figure 1.3(c). Two layer devices operating in this manner, are known as zener diodes and are used as low voltage reference sources. At higher reverse voltages, avalanche multiplication gives rise to a constant voltage characteristic. In the avalanche effect, a high speed electron dislodges an electron from an atom giving rise to a hole and a free electron, the action being cumulative.

## The tunnel diode

In the normal junction diode, there is a region between the two sections with very few free electrons or holes, which is known as the depletion layer. This region amounts to a potential barrier and to pass current, sufficient external voltage must be applied to overcome it. If the junction between the P and N sections is heavily

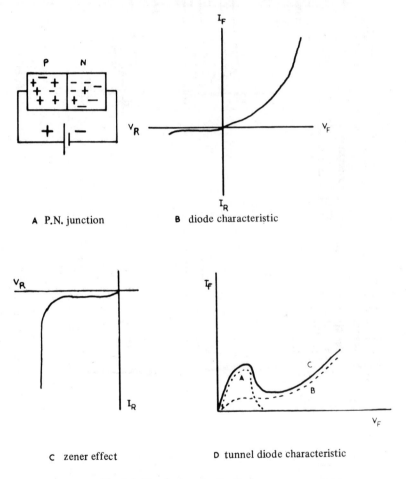

A  P.N. junction

B  diode characteristic

C  zener effect

D  tunnel diode characteristic

*Fig. 1.3.* The semiconductor diode.

doped with impurities a very narrow depletion layer results. Current carriers (positive holes or electrons) can now reach the other side of the potential barrier by tunnelling through it, this type of two layer semiconductor being called a tunnel diode. Its characteristic is shown as figure 1.3(d).

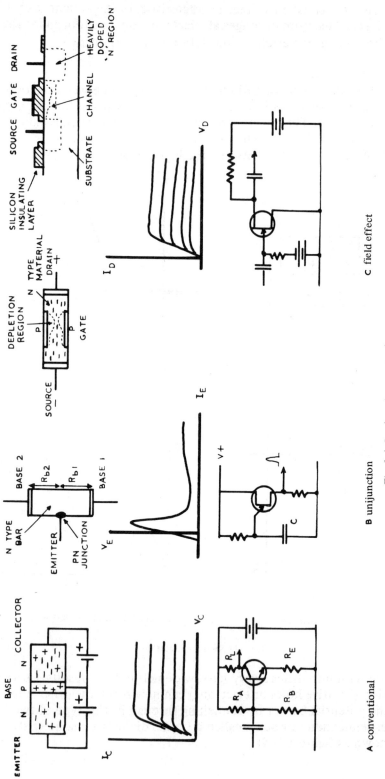

Fig. 1.4. Semiconductor transistors.

At small forward bias voltages we have the tunnelling current shown by curve A. This current however, disappears as the voltage is increased, and with the conventional diode current given by curve B, we obtain the composite characteristic of curve C. The tunnel diode may be used in many ways, examples being its use as an amplifier, oscillator and high speed electronic switch.

In the conventional diode, current flow depends on the ability of carriers to surmount potential barriers and is temperature dependent. The tunnel diode characteristic however is extremely stable with temperature, as the current flow is primarily a function of the tunnelling effect, which is of an electro-magnetic nature and not temperature dependent.

### The transistor

The transistor shown in figure 1.4(a) is a three layer semiconductor device and it may be of the N.P.N. or P.N.P. form. In the left hand junction (forward biased) the outer section is called the emitter and in the right hand junction (reverse biased) is called the collector. The very thin centre section has much lower conductivity than the outer sections and is called the base.

Nearly all the current carriers from the emitter pass through the base to the collector, as the carriers prefer to travel across the base region rather than take the high resistance path down the base and back to the first battery. An input voltage that increases the forward bias of the emitter, with respect to the base, increases the current carrier flow. A decrease of carrier flow is produced if the forward bias is decreased. The ratio $\frac{\text{change in collector current}}{\text{change in emitter current}}$ is known as the current gain alpha, for this common base configuration.

A positive signal applied to the base of the N.P.N. transistor amplifier circuit shown, will assist the forward bias voltage and the collector current will increase. This transistor is connected in the common emitter configuration and may be compared with the triode valve amplifier. The current gain given by the ratio $\frac{\text{change in collector current}}{\text{change in base current}}$ is known as beta.

For satisfactory operation it is necessary to temperature stabilise the transistor amplifier and one method is shown in the circuit of figure 1.4(a). In this circuit, if the collector current rises with temperature, the voltage drop across the emitter resistance $R_E$ is increased and the base is made more negative with respect to the emitter. The forward bias is thus decreased and therefore also the base current. This causes the collector current to drop and thermal runaway which would destroy the transistor is prevented, and the operating point is stabilised.

### The unijunction transistor

The unijunction transistor shown in figure 1.4(b), consists of a 'bar' pellet of N type silicon to which an emitter connection is made with P type material. The resistance of the 'bar' between each of the base connections is several thousand ohms, and the emitter connection has a potential which is proportional to the ratio of $R_{b1}$ to $R_{b2}$. The ratio of this voltage to the inter-base voltage $V_{bb}$ is a constant, known as the intrinsic stand off ratio (n). If the emitter voltage is below n times the inter-base voltage, the emitter junction is reverse biased and only a small current flows. If the voltage at the emitter is greater than n times the inter-base voltage, then the junction becomes forward biased and the emitter current increases considerably. Holes are injected into the 'bar' and drawn to the B1 connection increasing the emitter current flow. This increase is steep, as electrons are attracted to the holes, and the 'bar' resistance between the emitter and B1 drops rapidly.

The unijunction transistor may be used for a wide range of timing and oscillator circuits and a simple pulse generator circuit is shown. At the start of the operating cycle the emitter is reverse biased. As the capacitor C is charged, the voltage across it rises until the peak point voltage $V_p$ is reached, when the emitter becomes forward biased. The capacitor then discharges through the emitter to generate the output pulse.

### The field effect transistor

A relatively new device known as the field-effect transistor is shown in figure 1.4(c). If the P.N. junction between the gate and the N type bar is reverse biased, the majority carriers are repelled away from the junction and a depletion region is formed. By applying a variable negative voltage to the gate, we can vary the space charge distribution in the N type bar and control the flow of free electrons from the source to the drain connection. The path through the bar is known as the channel and can conduct an appreciable amount of current.

Since the gate-channel diode is reverse biased, little gate current flows and the field effect transistor thus gives large current and power gains. N channel field effect transistors are usually operated with the drain positive with respect to the source and the gate negative. P channel field-effect transistors operate with opposite voltage polarities.

The characteristic curve obtained is similar to that of the pentode valve, so that the field effect transistor may be used for a wide variety of applications. Important properties are its high input resistance (due to the reverse biased gate-channel junction) and low noise level, enabling the field-effect transistor to be used in amplifiers for the measurement of very small currents.

The type of field-effect transistor described above does not have the gate electrically insulated from the channel, and an input resistance of the order of $10^{10}$ ohms is obtained. A higher input resistance, up to about $10^{15}$ ohms, can be obtained in the metal oxide semiconductor transistor which has the gate electrode electrically isolated from the conducting channel. The construction of the metal oxide transistor is also shown in figure 1.4(c).

## P.N.P.N. devices

Another widely used semiconductor is the P.N.P.N. four layer silicon controlled rectifier or thyristor shown in figure 1.5. This device is used for power control, as a small change of the order of milliwatts at the gate terminal switches the transistor from a closed to an open state controlling a load of perhaps several kilowatts. The thyristor is normally biased well below the break over voltage, so that in this closed state, the current is very small. By injecting a current into the gate or trigger electrode, the operating mode is changed to the high conduction region and the trigger electrode no longer has any control. To turn off the device we must reduce the main current below the holding value and this can be done by applying a negative voltage to the anode. The controlled rectifier can thus be considered the semiconductor equivalent of the thyratron valve.

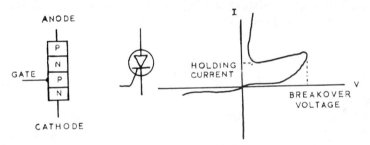

*Fig. 1.5.* The P.N.P.N. semiconductor.

## INTEGRATED CIRCUITS

### Thin film circuits

Conventional transistors and the field effect transistor are derived from single crystals suitably doped. Thin film transistors however use an extremely thin layer of polycrystalline N type material deposited on an insulated substrate. An insulated gate connection controls the current flow through the N type material, and a characteristic similar to the field-effect transistor is obtained.

Micro-miniaturization of electronic circuits has proceeded rapidly in the last few years, especially in the computer field. Integrated circuits of amplifiers and logic circuits are readily available, the circuit elements being constructed within a surrounding substrate. These integrated circuits may be divided into two types, (1) thin film circuits and (2) semiconductor circuits.

The thin film integrated circuit has the circuit elements deposited on a suitable substrate such as borosilicate glass or a ceramic. The usual deposition methods are vacuum deposition and sputtering. Passive elements such as resistors and capacitors can be deposited quite accurately but the production of transistors and diodes by similar methods is more difficult, and thin film integrated circuits may use physically small resin encapsulated transistors bonded to the thin film.

**Semiconductor integrated circuits**

In the semiconductor integrated circuit, a very pure silicon crystal is grown and later cut into slices which are then cleaned and smoothed. Both passive and active elements are produced by multiple diffusion on to one or more silicon chips.

Integrated circuits give increased reliability, reduce costs and enable a considerable reduction in size to be made for a given circuit function. Using an integrated circuit, the function of many transistors, diodes and resistors can be accommodated in the same space as one conventional transistor.

## AMPLIFIERS

**Types of amplifier**

The conventional amplifier is given by the valve common cathode or transistor common emitter connection. The voltage gain is greater than unity and there is a phase or polarity reversal between the input and output voltages. In the valve common grid and transistor common base connections, there is no voltage phase reversal between the input and output. A voltage gain of slightly less than unity is given by the valve common anode and transistor common collector connections. These circuits, known as the cathode follower and emitter follower, are useful for resistance or impedance matching between a high or moderate input resistance and a low output resistance.

In the low frequency a.c. amplifier, resistance capacitance coupling between stages is normally used. Transformers may be utilised but have poor frequency response.

Amplifiers may also use a parallel tuned circuit as the anode or collector load, when only one or a narrow band of frequencies is amplified. At the resonant frequency of the tuned circuit, the current is a minimum and the impedance a maximum. The maximum amplifier gain therefore occurs at the resonant frequency. The frequency at which resonance occurs is given by $f = \dfrac{1}{2\pi\sqrt{L.C.}}$ if the ohmic resistance of the inductance L is ignored.

The pass band of the amplifier is taken at the − 3dB points (i.e. where the gain falls to 0.707 of the maximum value) and depends on the coupling method and the Q factor of the inductance. The Q magnification factor of the inductance coil equals $\dfrac{2\pi fL}{R}$, where L is the coil inductance (Henries) and R its resistance (ohms). In high Q circuit the selectivity is high, the amplifier having a very narrow pass band. A broader pass band is obtained with a low Q inductance.

### Power output stages

Transformers are used in power amplifiers for matching the impedance between the load and the amplifier. In the single ended amplifier, the primary of the transformer is in series with the valve anode or transistor collector. To develop more power, the push-pull circuit shown in figure 1.6 is used.

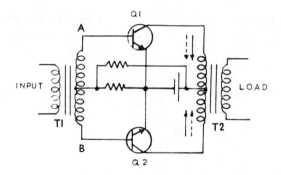

*Fig. 1.6.* Basic Push-Pull amplifier.

At any instant, the ends A and B of the transformer T1 secondary winding are 180° electrically out of phase. If end A of transformer T1 is positive and end B negative, transistor Q1 base voltage is then positive going, and Q2 base voltage negative going. This results in Q1 conducting more than Q2. When the transformer polarity reverses, Q2 conducts more than Q1. The heavy arrow in the diagram represents the heavily conducting transistor current and the dotted arrow the lightly conducting transistor current. These currents are additive, and flow through the output transformer T2 primary so that they induce voltages in the secondary which are in the same direction. The secondary output voltage is then an amplified reproduction of the signal. This type of circuit is typically used for driving motors in control systems, and the quiescent bias voltages are usually such that current only flows for one half cycle of the input voltage (class B operation).

**Direct coupled or zero frequency amplifiers**

If we wish to measure and amplify d.c. voltage changes, then capacitors cannot be used for inter-stage coupling and special methods are necessary, as the 'high' voltage at one valve anode cannot normally be connected straight on to the grid of the following

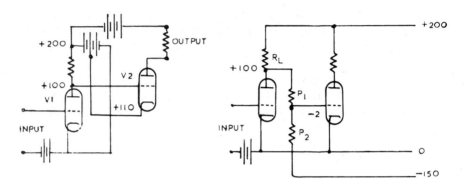

A  use of batteries for coupling      B  positive and negative supply rails

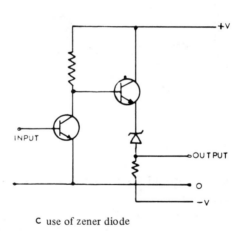

C  use of zener diode

Fig. 1.7. Direct coupled amplifiers.

valve. Coupling batteries, potential divider circuits and cold cathode diodes are used to establish the correct quiescent operating points and two methods used in valve circuits are shown in figure 1.7. Similar methods may be used in transistor circuits and one circuit using a zener diode is also illustrated. Direct coupling between two transistors is possible if the collector voltage of the first stage is low enough to give the right bias condition for the second.

One of the problems encountered when amplifying d.c. voltages, is that changes in the H.T. supply or valve filament heater voltages give rise to spurious signals which may be amplified giving rise to errors at the output. This is known as drift. These changes are prevented from giving rise to errors in a.c. amplifiers by the coupling capacitor. Transistor amplifiers are normally of the common emitter type and a change in the d.c. operating point of any stage immediately affects the succeeding stages, the collector current changing by beta times the base input current change. Transistor parameters which vary with temperature are (1) the base-emitter voltage $V_{BE}$ (2) the d.c. current gain $h_{FE}$ and (3) the collector leakage current

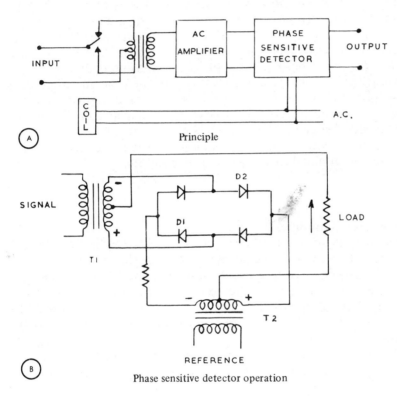

*Fig. 1.8.* The chopper amplifier.

$I_{CO}$. In transistors, temperature variation is the most important cause of drift, while in valves the effect of ambient temperature change is less important than changes in the supply voltage and valve parameters.

Some methods used to overcome drift are (1) the use of temperature sensitive elements to counteract the increase of transistor collector current with temperature (2) the use of emitter coupled transistor pairs or common cathode connected valves (3) the use of regulated power supplies and (4) conversion of the d.c. signal to a.c. which is amplified and then converted back to d.c. The latter method uses a chopper amplifier (figure 1.8(a)) which uses an electro mechanical chopper or vibrator for d.c. to a.c. conversion, an a.c. amplifier, and a phase sensitive rectifier to convert the amplifier a.c. output back to d.c. All these techniques are extensively used and circuits employed in various instruments will be described in the following chapters.

The basic phase sensitive rectifier or detector is shown in figure 1.8(b). This circuit is of the full wave type and uses four diodes which are controlled by the a.c. reference voltage input. A d.c. voltage is developed across the load proportional to the amplitude of the signal. When end A of transformer T1 is positive, diode D1 conducts, and current flows in the load as indicated by the arrow. On the next half cycle the transformer secondary polarities reverse and diode D2 conducts, current flow through the load being in the same direction as before. If the phase relationship between the reference and signal wave form is reversed, then the direction of the load current also reverses. The phase sensitive rectifier thus gives a direct voltage output, the polarity of which depends on the phase relationship between reference and signal input. In the chopper amplifier the reference voltage and supply for the coil of the vibrator are usually obtained from the a.c. mains supply.

**Electrometer amplifiers**

In many instruments, the output from the detecting device (e.g. a photocell or ionisation chamber) is very small, and in order to develop sufficient voltage to feed into an amplifier it is necessary to use a very high ohmic value resistor. This resistance may be as large as 2,000,000,000 ohms and in consequence it is necessary to use a special valve known as an electrometer as the amplifier first stage.

Electrometer valves are specially made of selected low voltage types that have a high input resistance and very low grid currents (which may flow in or out of the valve grid).

Some of the sources of grid current are (1) collection at the grid of emitted electrons from the cathode (reduced by biasing the grid negative with respect to the cathode), (2) leakage over the glass or insulation (reduced by guard insulation rings or by operating the valve in an enclosed dry atmosphere), and (3) photo emission of electrons from the grid under action of light from external sources or from the valve heaters. External light can be prevented from falling on the grid by coating the tube with black paint and heater glow reduced by operating at low voltages.

When measuring the voltage drop across a very high ohmic value resistance, the insulation resistance of the fixtures, wiring and electrometer input valve must be many times greater than the resistance itself. If this is not the case, then the high resistance voltage source will be partially short circuited, and true readings will not be obtained due to the leakage current which will flow. Insulators suitable for electrometer work are Teflon, Polystyrene, Kel F, Polyethylene, Glass or Ceramic.

Short stiff unsupported wire is the best method of connection between the live side of the source and the electrometer input. Where high ohmic value resistors are used they are usually mounted on Teflon standoffs so that the insulation resistance to ground or earth is of the order of $10^{15}$ ohms. When the E.M.F. source is remote from the electrometer, polyethylene- or teflon-insulated co-axial cable is used. An example of this is to be found in the measurement of pH, where the insulation of the leads should be of the order of 100,000 Megohms.

The high resistance circuitry should be well shielded or screened, as pickup from stray electric fields will give rise to spurious potentials.

In order to overcome the effects of grid current, electrometer amplifiers may utilise vibrating capacitor inputs in which the high resistance source is not directly connected to the electrometer input. In this technique, the input d.c. potential gives rise to a variable potential across the plates of the vibrating capacitor. This variable component is fed to the amplifier through d.c. blocking capacitors.

The metal oxide transistor with its very high input resistance may also be used as an electrometer amplifier.

### Feedback and stability

In order to stabilise the gain of an amplifier, negative feedback is employed. A block diagram of a feedback amplifier is shown as

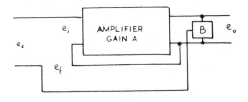

*Fig. 1.9. Amplifier feedback.*

figure 1.9 and the amplifier gain without feedback is given by $A = \dfrac{e_o}{e_i}$. If a fraction B of the output voltage is fed back to the amplifier input, so that the phase of the feedback decreases the effective input, the amplifier input signal is given by es = ei − ef = ei − Beo. The amplifier gain with feedback is reduced and is given by

$$A_f = \frac{eo}{ei - Beo} = \frac{A}{1 - AB}.$$

The fraction B of the output voltage may be obtained by a resistance voltage divider circuit, and it can thus be seen that the gain of the amplifier depends on the stability of these resistors, and is independent of variations in valve or transistor characteristics and operating voltages.

Both current and voltage feedback networks are used and in the simple transistor amplifier shown in figure 1.4(a) we have current feedback stabilisation. The feedback ratio B is given by $\frac{RE}{RL}$.

It should be noted that while negative feedback stabilises an amplifier, gain and drift are reduced proportionately so that there is no improvement in the gain/drift ratio.

**The long tailed pair circuit**

The emitter coupled (or valve cathode coupled) amplifier gives satisfactory stability for a wide range of applications. This type of amplifier is shown in figure 1.10, and ideally the two transistors should have identical parameters. The collector currents must be equal and potentiometer RV1 provides adjustment for any unbalance of $V_{BE}$ and $h_{FE}$ (gain).

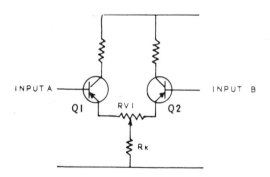

*Fig. 1.10.* Emitter coupled amplifier.

If a negative going signal is applied to the base of transistor Q1, then the current through it increases. Due to the common emitter coupling, the base-emitter potential of Q2 is reduced and therefore Q2 collector current decreases. If resistance $R_k$ is high, the total current through the common emitter resistance tends to remain constant.

This circuit is known as a long tailed pair and variations in the quiescent voltages are largely cancelled out and stable operation results. The input signal causes the current in one half of the amplifier to increase and the current in the other half to decrease, so that a voltage difference is produced between the two transistor collectors. This circuit can be used to obtain two equal anti phase signals from a single input and is then known as a paraphase or phase splitting amplifier.

### Differential amplifiers

If inputs are applied to the bases of both transistors and only one output is used we can obtain an output signal which depends on the difference between the two inputs. Used in this manner the circuit is called a differential amplifier and an output is obtained if the inputs are of opposite phase or are of unequal amplitudes. The amplifier will also amplify single ended inputs (e.g. Q1 base input e1 and Q2 base input zero). Under ideal conditions, common in-phase inputs applied simultaneously to both inputs should produce no output. The ratio of the gain obtained with differential inputs applied, to that of the gain obtained with both inputs tied together (common), is known as the common mode rejection factor.

In order to reduce the common mode gain and improve the rejection ratio the common emitter resistance should be as high as possible. This however, reduces the operating current and the voltage must be increased. An alternative is to use a third transistor as a constant current source, the effective resistance then obtained being several megohms.

### Operational amplifiers

The operational amplifier was originally used in analogue computing circuits, and may be considered as an amplifier with feedback circuits such that the output is a mathematical function of the input. The principle is shown in figure 1.11, where if ZF is capacitive and Z1 resistive, the output is proportional to the integral of the input voltage. The integrating capacitor must have low leakage and low dielectric absorption so that small capacitors and large input resistances are used. This has the disadvantage of a low feedback factor and gives rise to noise. If the impedances are reversed, so that ZF is resistive and Z1 capacitive, we have a differentiating circuit. With a diode or transistor in the feedback loop the output is proportional to the logarithm of the input.

Since the change at the base of the input transistor is small, it can to a first approximation be assumed to be zero, and this simplified virtual earth concept of the operational amplifier is also illustrated.

Generally speaking, operational amplifiers have high d.c. voltage gain (up to $10^9$), a high input impedance, and low input d.c. offset and drift with time and temperature. Offset is the d.c. output signal obtained with zero input signal, and it may be compensated

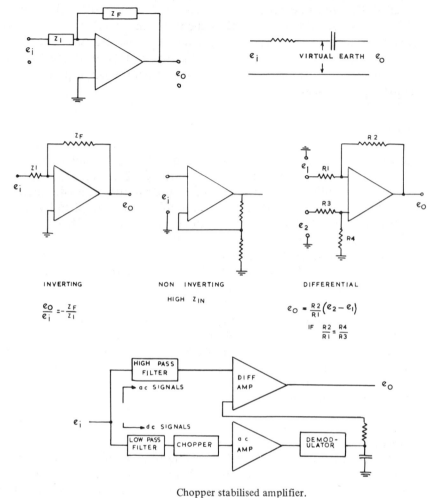

Chopper stabilised amplifier.
*Fig. 1.11.* Operational amplifier configurations.

by applying an equal and opposite signal at the input summing junction. The input offset is the input required to zero the d.c. component of the output with zero input signal.

The three main configurations in which operational amplifiers are used: inverting, non-inverting and differential, are also illustrated in figure 1.11. Chopper stabilised operational amplifiers which incorporate an a.c. amplifier, chopper, demodulator and a differential amplifier, are used to obtain improved drift characteristics and long term stability. The chopper amplifier corrects the drift of the main amplifier channel which is effectively divided by the gain of the chopper amplifier. All of these units are available as plug in printed circuit modules. Operational amplifiers with field effect transistor inputs are available and have a very high input impedance.

### Noise

When measuring low level voltages, spurious signals (noise) may be troublesome. These unwanted signals may be due to hum pickup although this can be eliminated by using d.c. for valve heaters, resistance capacitance filters in the H.T. supply lines and physical isolation from alternating magnetic fields. Spurious signals may also arise as a result of mechanical vibration. This is known as microphonic noise. Pickup from external sources may be minimised by proper shielding and avoiding multiple earthing points, as potential differences may exist between them.

Noise in a circuit may be due to inherent noise in the circuit components. Typical sources are the thermal motion of electrons in a load resistor, shot effect, flicker effect and grid current in valves. Shot noise is due to random variations in the rate of arrival of electrons at a valve anode. The flicker effect is modulation of the cathode current by sudden changes in potential difference across the boundary layer of the cathode.

Flicker noise not only occurs in valves but also in semiconductors and composition resistors. In semiconductors, majority carriers become trapped in lattice imperfections of the material and influence the rate of generation and recombination of charged carriers. Surface contamination of a semiconductor will cause leakage currents and give rise to noise. In resistances, variation in the contact resistance between granules generates flicker noise.

Multigrid valves are generally noisier than triodes because the current is divided between the various electrodes — partition noise. With negative feedback, noise such as power supply ripple (hum), is reduced if it enters the amplifier circuit after the first stage.

Power amplifying devices have a noise spectrum that varies as the inverse of the frequency, but when amplifying at zero frequency (d.c.) attempting to improve the signal to noise ratio by reducing the band width of the amplifier is useless. If however the signal is chopped at a frequency fo, and this is applied to the signal input of a phase sensitive rectifier (the reference input also being at frequency fo), it is possible to amplify with the introduction of a minimum amount of noise. To avoid the region of noise the frequency fo should ideally be above about 100 Hz for valves and 1,000 Hz for transistors.

## OSCILLATORS

### Principle of operation

When the feedback voltage in an amplifier is in-phase with the input voltage we have positive feedback. The positive feedback action is cumulative and the circuit will oscillate. This is the principle of operation of oscillators and multivibrator circuits, the multivibrator consisting basically of a two stage amplifier with a large amount of positive feedback.

### Multivibrator circuits

There is a large family of multivibrator circuits or relaxation oscillators as they are sometimes called. The astable type, switches from the on to the off state at regular time intervals without any external triggering, and may be used as a square wave generator, electronic switch, or as an electronic gate which passes signals to the output only during a desired time interval. The bistable circuit has one stage either on or off until an external triggering pulse reverses conditions, and is extensively used in binary counting circuits as one output pulse is obtained for every two input pulses. A variation of the circuit is the cathode coupled binary or Schmitt trigger which gives a constant amplitude square wave output when the input triggering pulse exceeds a certain threshold value.

When a triggering pulse is applied to a monostable or one shot circuit, the on and off stages reverse and after a time, determined by the circuit R.C. time constants, revert to their original state. This circuit is often used for pulse equalisation in counting circuits. which are critical with respect to the size and shape of pulses. These pulses trigger a one shot circuit which gives a uniform output pulse.

The operation of some of these pulse circuits, is described in chapter 5.

### Sinusoidal oscillator circuits

It is difficult in practice to obtain sustained sinusoidal oscillations over a single valve or transistor stage unless a phase shift network is used. This network provides a 180° phase shift between a valve anode and the grid so that positive feedback is maintained. An oscillator of this type, known as a resistance capacitance coupled or phase shift oscillator, is shown in figure 1.12(a) and has a feedback network consisting of three resistance capacitance sections.

The feedback may also be obtained by the mutual coupling between two coils or by capacitive coupling. Circuits of this type are known as tuned circuit oscillators and examples are the Colpitts circuit (capacitance coupling), and the Hartley circuit (mutual coupling between the coils) which is shown in figure 1.12(b).

Another type of oscillator uses a Wein bridge circuit in which a two stage amplifier is connected between the output and input of the bridge. This is shown in figure 1.12(c) and the R1.C1.R4.C4. network may be regarded as giving positive feedback, while R2 and R3 simultaneously give negative feedback to limit the amplifier

# HIGH FREQUENCY TECHNIQUES

gain to the required value. In order that oscillation is possible, the amplifier gain must not be less than the attenuation of the bridge network.

For very stable oscillations, quartz piezo-electric crystals which act as tuned circuits are used. These crystals have an extremely high Q value, several hundred times that which is obtainable with a mag-

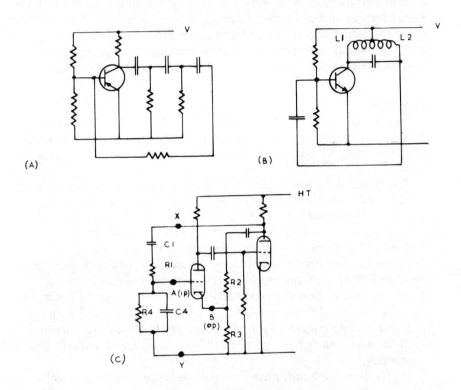

*Fig. 1.12.* Basic oscillator circuits.

netically coupled oscillator such as the Hartley type. Up to about 300 MHz, oscillators of the magnetic or capacitance coupled type may be used to generate metre waves. Above this frequency it is necessary to use microwave devices such as the klystron which generate centimetre waves.

## HIGH FREQUENCY TECHNIQUES

In addition to radio and audio frequency oscillators, a number of radio techniques are used in electro-chemical, nuclear magnetic, and electron spin resonance equipment. Some of these techniques are outlined below.

## Frequency changing

A twin triode or a valve with two grids may be used for frequency changing or mixing. With a signal voltage of radio frequency f1 on grid 1 and a voltage of frequency f2 generated by a local oscillator on grid 2, a difference frequency known as the intermediate frequency (I.F) is produced at the output. In order to prevent interaction, extra screen grids were introduced, giving rise to the Hexode valve. This valve has signal, screen, oscillator, screen, and suppressor grids between the cathode and anode. Heptode valves which contain a triode oscillator section in the same envelope as the mixer may also be used.

## Detection

Diodes may be used for demodulation or detection, the positive halves only of the input radio or audio frequency signal appearing at the output. At microwave frequencies special crystal detectors are used.

The phase sensitive (lock in) detector is a circuit used extensively in radio and microwave spectroscopy. This circuit has two inputs, one from the signal source and another from a reference oscillator. The reference is usually of sinusoidal wave form and the signal a voltage of the same frequency. The output is a d.c. voltage proportional to the amplitude of the in-phase component of the signal. Response of the circuit to quadrature components and noise is small and in addition the phase sensitive detector may be preceeded by a narrow bandwidth amplifier. A practical circuit may utilise a phase shift amplifier between the oscillator input and the mixer section, where the phase adjusted reference is mixed with the signal frequency.

At the very high radio frequencies, specially constructed valves with anodes and grids in disc form to overcome inter-electrode capacitances are used. For operation at microwave frequencies (centimetre wave lengths) metallic tubes known as waveguides are necessary for transmission, and special valves known as klystrons and magnetrons are used.

## The klystron

In microwave spectrometers, the klystron is normally used as an oscillator and its operating principle is shown in figure 1.13(a). The beam of electrons from the heated cathode passes through the buncher to the collector electrode. The buncher and catcher are resonant cavities which resonate when an electron beam passes their opening. In effect they act as tuned circuits.

The buncher velocity modulates the electron beam, the electrons being accelerated, retarded or unaffected depending on the potential of the buncher grid furthest from the cathode. When the alternating potential is positive going, electrons are accelerated but as the positive potential is reduced to zero, the velocity of the electrons is decreased until they have the same velocity they possessed before reaching the buncher. With the potential negative going, the electrons are retarded and leave the buncher at a lower velocity than when they arrive. The radio frequency electric field from the first resonator thus velocity modulates the electrons.

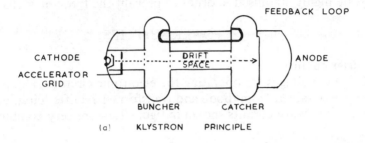

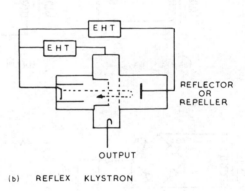

Fig. 1.13. The Klystron.

After leaving the buncher, the electrons enter a field free region known as the drift space, where the quicker electrons overtake the slower ones, resulting in the electrons being concentrated into bunches. The bunched electrons in passing across the grids of the catcher resonator, excite oscillations if the frequency of the bunches matches the resonant frequency of the cavity.

If the system is designed so that the amplitude of oscillation in the second cavity is greater than that in the first, amplification is obtained. If a feedback loop is provided, with proper phasing and tuning, the device becomes an oscillator. The cavities are tuned by mechanical deformation of the walls or by using plungers.

Klystrons are usually of the reflex type shown in figure 1.13(b). After passing the resonant cavity grids, the electrons are retarded and repelled by the repeller electrode which has a negative potential applied to it. The electrons which are now bunched, return to the region of the cavity which acts as a catcher. In the reflex klystron, the problem of tuning two cavities to the same frequency is eliminated and fine tuning may be obtained by altering the voltage on the repeller. Most klystrons require several thousand volts HT which must be highly stabilised in order to prevent the frequency changing.

## POWER SUPPLIES

### Rectifier circuits

The d.c. voltages for operating the electronic circuits in an instrument are obtained from diode and transformer-rectifier circuits. The half and full wave circuits shown in figure 1.14 are very common

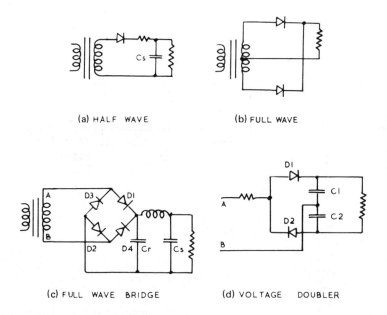

Fig. 1.14. Basic rectifier circuits.

and the resulting wave form may be smoothed by a resistance capacitance network so that the amount of ripple superimposed on the d.c. voltage level is reduced to a minimum.

The half wave circuit shown has a capacitor input filter for smoothing. Capacitor $C_s$ is charged to the maximum value of the applied volts on the positive half cycle of the a.c. input voltage wave form. When the voltage falls and during the negative half cycle of the input, the capacitor discharges into the load. Disadvantages of the half wave circuit are its low efficiency and the large low frequency ripple. They are overcome in the full wave rectifier circuit which can be of the centre tapped transformer or bridge type.

In the centre tap circuit, the rectifiers operate in push-pull on alternative half cycles. With the bridge rectifier arrangement, when end A of the transformer is positive, diodes 1 and 2 conduct. Diodes 3 and 4 conduct when end B is positive.

## Smoothing

A $\pi$ section filter is shown in figure 1.14(c) and is used where a high degree of smoothing is required. Capacitor $C_R$ is a reservoir capacitor and capacitor $C_S$ provides additional smoothing. The choke is sometimes replaced by a resistance of a few thousand ohms. A disadvantage of this filter circuit is the voltage drop produced across the choke or resistor.

## The voltage multiplier

Instruments such as geiger radiation counters require a low current high voltage d.c. supply, and to obtain this it is often preferable to use a voltage multiplier circuit instead of a high voltage winding on a transformer.

A simple voltage doubler is shown as figure 1.14(d). When the a.c. input at A is positive, current flows through diode D1 and capacitor C1 is charged. Capacitor C2 is charged when input B is positive. The output is taken across both capacitors and the d.c. voltage is nearly twice the value of the peak a.c. voltage applied between A and B. Using further rectifiers and capacitors, voltage tripler and quadruplers may be built up.

## Stabilised supplies

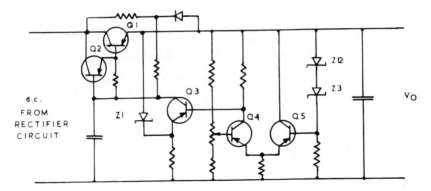

*Fig. 1.15.* Stabilised d.c. voltage supply.

In the majority of instruments with which we are concerned it is necessary to stabilise the supply to the electronic circuits. This ensures that the internal voltages remain constant regardless of variations in load and the a.c. mains supply.

A stabilised voltage supply circuit is shown in figure 1.15. In this circuit, a fraction of the output voltage is compared with a stable reference voltage obtained from the zener diodes Z2 and Z3. The comparison circuit is a differential amplifier formed by transistors Q4 and Q5. If the output voltage $V_o$ increases, an output is obtained from the differential amplifier and a negative going signal is applied to transistor Q3 base. The collector of transistor Q3 controls the base of Q2 which in turn controls Q1. The base current of Q1 is reduced and a compensating volts drop is produced across this series controlling transistor. The output voltage is then regulated to the desired value. Transistors Q1 and Q2 are connected as a Darlington pair (cascaded emitter followers).

Stabilised power supplies may be designed to regulate the voltage applied to a circuit or to maintain the current in a load at a constant level. Constant current power supplies are used in spectrophotometry to control the arc current of deuterium or hydrogen discharge lamps.

## Books for Further Reading

### General
G. C. WARE. Basic electronics for biologists. J. & A. Churchill Ltd.
E. E. ZEPLER & S. W. PUNNETT. Electronic circuit techniques. Blackie.
E. E. ZEPLER & S. W. PUNNETT. Electronic devices & networks. Blackie.
E. WILLIAMS. Thermionic valve circuits. Sir Isaac Pitman & Sons Ltd.
A. H. SEIDMAN & S. L. MARSHALL. Semiconductor fundamentals. John Wiley & Sons Inc.
Reference manual of transistor circuits. Mullard Ltd.
H. JACOBOWITZ. Electronic computers made simple. W. H. Allen.
F. J. M. FARLEY. Elements of pulse circuits. Methuen & Co. Ltd.
K. J. DEAN. An introduction to counting techniques and transistor circuit logic. Chapman & Hall Ltd.

### Electronic Instrumentation
P. E. K. DONALDSON. Electronic apparatus for biological research. Butterworth Publications.
D. DOBOS. Electronic electrochemical measuring instruments. Terra. Budapest.
H. E. SOISSON. Electronic measuring instruments. McGraw-Hill Book Co. Inc.
K. J. DEAN. Digital instruments. Chapman & Hall Ltd.
C. G. CANNON. Electronics for Spectroscopists. Hilger & Watts Ltd.
D. J. INGRAM. Spectroscopy at radio & microwave frequencies. Butterworth Publications.
T. H. WILMSHURST. Electron spin resonance spectrometers. Hilger & Watts Ltd.

# 2 Electrochemical Instruments

## pH ELECTRODES

**The pH-EMF equation**

The pH value of a solution, which may be defined as minus the logarithm of the hydrogen ion concentration, is usually measured by a glass electrode with a reference electrode to complete the electrical circuit (38) (74) (9). With the electrodes immersed in the solution, the cell EMF is given by $E = E_o + (0.1984T \times pH)$ where E is the measured EMF in millivolts, $E_o$ a zero term and T the absolute temperature. The equation shows that in addition to the pH proportional EMF we also have an almost constant standing voltage $E_o$, which depends on electrode constants and the glass membrane composition and construction. The value of this standing voltage changes slowly with time and so necessitates standardisation against a buffer of known pH at frequent intervals.

The pH value of a solution at which the measuring assembly gives zero EMF is known as the check or zero point value, and figures are quoted for glass electrodes with reference to a standard calomel type electrode. Glass electrodes giving a zero point of 2, 6 and 7 pH are usual and the pH meter circuit should be set up to correspond to the electrode measuring system in use. If the electrodes have a zero point of 2pH then the electronic measuring circuit must be adjusted so that the meter reads a value of 2pH with zero input. This is achieved by applying a backing off voltage using the coarse buffer or zero controls located inside or at the rear of the instrument; the fine buffer control being set to its mid point. Should we now use electrodes with a zero point of 7pH it may not be possible to set the meter to the standardising buffer pH values without first re-adjusting the backing off voltage. This is done in the same manner as before, the meter being set to read 7pH with zero input. It is usual to short-circuit the glass and reference electrode terminals on the meter when carrying out this adjustment in order to ensure that there is no input to the electronic circuit. This may be done automatically by means of an input selector switch, as on E.I.L. pH meters, the input being short circuited with the switch in the check position.

## Glass electrodes

There is a wide variety of electrodes of varying sizes and shapes which cater for different applications. These range from spear shapes for immersion into semi solids to flat membranes for paper or skin measurements. The type of glass used in the electrode construction also depends on the application and examples are shown in table 2.1. The high resistance of the glass electrode membrane necessitates the use of an electronic measuring circuit although low resistance electrodes may be used with a simple potentiometer circuit.

*Standard and General Purpose Glass Electrodes*

| Type | Construction | pH range | Temperature range | Zero point |
|---|---|---|---|---|
| E.I.L. GG23 | Mac-Innes-Dole glass | 1-9.5 | 10-45°C | 2 pH |
| GHS23 | GHS glass | 0-14 | 10-140°C | 2 pH |
| Pye Ingold 201 | glass resistance 200-300 MΩ at 20°C | 0-13 | 0-70°C | 2 pH |
| Pye 11126 | Lithium glass | 0-13.5 | 0-100°C | |
| Radiometer G202B | B type glass. Only minor salt correction above pH 13 | 0-14 | 20-60°C | 6.5 pH |
| Beckman A7LB | | 0-14 | 10-140°C | 7 pH |
| A2LB | | 0-14 | 10-140°C | 2 pH |

*Low Temperature Glass Electrodes*

| Type | Construction | pH range | Temperature range | Zero point |
|---|---|---|---|---|
| E.I.L. GC23 | Low resistivity glass BH 25 | 0-11 | 0-50°C | 2 pH |
| Beckman X21C | Low resistivity glass | 0-11 | 0-50°C | 2 pH |
| Pye-Ingold LOT201 | Glass resistance 30-50 MΩ | 0-12 | −10 to +70°C | 2 pH |

*High Temperature Glass Electrodes*

| Type | Construction | pH range | Temperature range | Zero point |
|---|---|---|---|---|
| E.I.L. GB33 | BH 15 glass | 0-13 | 50-140°C | 2 pH |
| Radiometer G202BH | B type glass | 0-14 | 40-120°C | 6.5 pH |

*Combined Electrodes*

| Type | Construction | pH range | Temperature range | Zero point |
|---|---|---|---|---|
| Radiometer GK202B | Inner glass electrode B type glass | 0-14 | 20-60°C | 6.5 pH |
| Beckman AD2LP | | 0-14 | 0-50°C | 2 pH |
| Pye-Ingold 401 | Glass resistance 100-200 MΩ | 0-13 | −5 to +70°C (intermittent to 100 °C) | 2 pH |

*Calomel Reference Electrodes*

| Type | Construction | pH range | Temperature range | Zero point |
|---|---|---|---|---|
| Beckman RLS | fused porous plug | − | −5 to +80°C | |
| Pye-Ingold 303 | Fused porous plug resistance 2 kΩ | − | 0-70°C | |
| E.I.L. RJ 23 | Heat sealed ceramic plug | − | 0-100°C | |

TABLE 2.1.
*CHARACTERISTICS OF SOME TYPICAL LABORATORY pH ELECTRODES*

## Metal electrodes

Metal electrodes are not much used for pH measurement, although the low resistance antimony electrode is useful when measuring strong fluoride solutions as they have a corrosive effect on the glass membrane. Platinum and gold spade electrodes are used for redox measurement and for quinhydrone pH work. The reference electrode for redox titration is also metallic and may consist of a small tungsten rod.

## Measurement errors

It should be noted that a small error occurs with the glass electrode when measuring highly alkaline solutions. This alkaline error is negligible with 0.1N sodium hydroxide at 25°C (pH 12.7) and about 0.2 pH with a normal solution at 25°C (pH 13.7). (80). For accurate and standard measurements free from salt errors (shift of pH caused by variations in salt content of a solution) the hydrogen electrode may be used. This electrode is difficult to make and consists basically of a noble metal with a fine adherent layer of noble metal. Hydrogen gas is used to saturate the electrode and its output is measured with a potentiometer as the electrode has a low resistance (64) (43).

## Specific ion electrodes

Specific ion electrodes, in which the potential is proportional to the number of specific ions present in the solution are now available. They are used with a special reference electrode and measurements can be made on a standard pH meter. The pH glass electrode may be considered a type of specific ion electrode as it responds to positive hydrogen ions. Sodium and Cationic electrodes have tips of specially formulated glass. Sodium specific ion electrodes are responsive to silver, hydrogen, sodium and potassium in descending order of sensitivity while Cationic electrodes are similarly responsive to hydrogen, silver, potassium, ammonium, sodium and lithium. The chloride ion electrode has a silver tip coated with silver chloride (10).

Many types of Cation sensitive membranes and silicon membranes packed with ion exchanger substances, which develop an E.M.F. depending selectively on the concentration of anions in solution, have been described. Any type of pH meter with a millivolt scale may be used with them (27) (91) (39).

## Carbon dioxide electrodes

The Severinghauss carbon dioxide electrode may be used with an electrometer amplifier. It consists basically of glass and calomel electrode elements, which measure the pH of a thin film of aqueous sodium bicarbonate, separated from the gas or liquid sample by a teflon membrane which is permeable to carbon dioxide molecules (96) (101). The change in pH of the electrolyte is proportional to the change in log $pCO_2$. If a pH meter is used with a linear recorder to continuously record $pCO_2$, an antilogarithmic circuit (e.g. diode with a logarithmic forward characteristic and an operational amplifier) may be used in order to establish a linear relationship between $pCO_2$ and the recorder response.

### pH reference electrodes

The pH reference electrode is basically a reference element separated from the test solution by means of a 'salt bridge'. The 'salt bridge' container should be filled about two thirds full with saturated potassium chloride and it is important that the level of the salt solution should always be above that of the solution in which the reference electrode is standing.

The thermal condition of the reference electrode affects the temperature compensation and stability, so the liquid junction tip only should be immersed in the solution under test. This allows the reference element to remain at ambient temperature even though the test solution temperature may be varying.

The salt bridge solution for the mercury calomel electrode is potassium chloride, although reference electrodes with different elements and salt solutions are available, and may incorporate mercurius and sodium sulphate where potassium or chloride ions are to be excluded.

Dual electrodes of concentric construction, with the salt solution surrounding the glass electrode, can also be obtained and are very useful when making pH measurements in small containers.

### Temperature effects

From figure 2.2(a) we see that the pH-EMF curve changes its slope with temperature and methods of compensation for this will be described when outlining typical pH meter circuits. Point A is known as the isopotential point, and is the pH value for a particular electrode pair at which the EMF is invariant with temperature, although the isopotential pH is only constant over a range of about 20°C. It can also be seen how the zero point of the pH-EMF curve changes with temperature and compensation for this is made by applying an electrical offset to the meter.

### Correct use of electrodes

For satisfactory operation it is essential to keep electrodes clean and they should be rinsed with distilled water between each sample measurement. For very accurate pH work however, it is better to rinse with a solution of the pH value next to be measured rather than with distilled water. Ultrasonics may also be used for electrode cleaning and this is especially useful in industrial applications. In the E.I.L. U28 system a lead zirconate transducer is used, the ultrasonic generator operating at 25kHz.

Care should be taken not to scratch the glass electrode, as a scratch or crack may give rise to a constant reading on the meter, regardless of the pH value of the solution. If sluggish response is observed this may be due to a coating on the glass electrode membrane which can be reactivated by soaking in 0.1N hydrochloric acid for 24 hours. This soaking should also be carried out on a new glass electrode and in both cases the electrode should be thoroughly rinsed with distilled water prior to use.

The electrical resistance of glass electrodes is typically between 50 and 500 megohms. It is therefore important, in order to avoid variable and spurious readings on the pH meter circuit, to ensure that leakage currents are reduced to a minimum. This may be done by (1) keeping the instrument case dry (internal heaters) and (2) ensuring that the insulation resistance to earth of the cable connected to the glass electrode is high.

Capacitance (electrostatic) effects may cause variable readings when the operator moves near the electrodes and this effect can be minimised by screening the input terminals of the electrodes by an earthed metal cover. Screened connecting cables are used for accurate work and eliminate erratic readings on the meter due to frictional electrostatic charges generated when the connecting cables are moved.

Due to the glass electrodes high resistance the electronic measuring circuit should have an input resistance of at least $10^4$ megohms if measurement is required to an accuracy of 1% (34). Errors arising in pH measurement, after standardisation at 25°C and subsequent use at lower temperatures are shown in a section of British Standard 2586. Errors due to grid current of the electrometer valve are also tabulated and it should be noted that these errors are not corrected for by temperature compensation devices (20).

Unstable extraneous asymmetry potentials may also exist across the glass electrode walls, and with a glass of composition 72% $SiO_2$, 8% CaO, 20% $Na_2O$, this potential is about 2 millivolts.

## pH METER CIRCUITS

**D.C. amplifier instruments**

A simple and inexpensive type of measuring circuit, used in the Beckman 72 pH meter, is shown in figure 2.1 where the EMF from the electrodes is applied to the control grid of valve V1. The potentiometer control and valve V2 provide a means of compensating for variations in the electrode standing potential, and enable the meter to be set at the correct position for standardising. The circuit may be considered as a bridge whereby any drift in the V1 arm is compensated by a change in the V2 arm. The two amplifier sections are

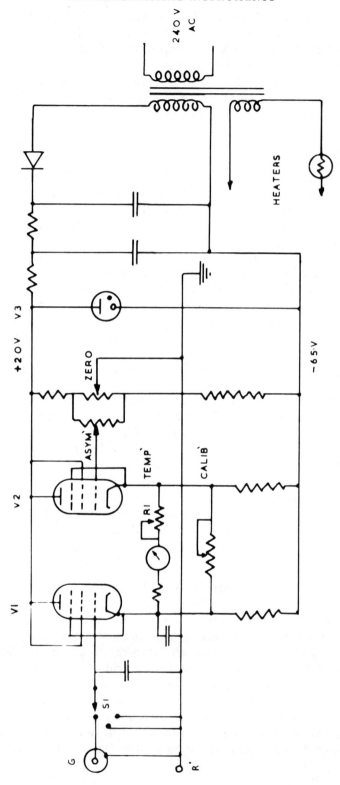

Fig. 2.1. Beckman 72 pH meter.

so connected that signals common to both, such as changes in the H.T. or heater supply, affect them equally, the changing cathode voltages by the same amount so the voltage between the cathodes remains unchanged. Signals applied asymmetrically to the inputs cause the cathode voltages to be at different levels so that an output signal is obtained.

As the electrode potential per pH unit of a solution changes approximately at the rate of 0.2mV per 1°C, compensation is necessary to vary the sensitivity of the meter circuit according to the temperature. The setting up procedure is then, first, ascertain the temperature of the solution and set the temperature calibrated compensation resistance R1 to the measured value. Next, immerse the electrodes in a buffer solution of known pH and adjust the potentiometer control until the meter reading corresponds to the buffer pH. The instrument is now calibrated and ready for use. The meter calibration and the zero control are internal screwdriver adjustments and are not used in the pH measuring procedure.

One of the difficulties associated with pH measurement is drift in the d.c. amplifier, and to overcome this problem cold cathode tubes may be used to stabilise the H.T. voltage and so minimize errors due to voltage variation. Some circuits also use barretters or transistors to stabilise the valve heater supply.

The barretter is a hydrogen filled tube with an iron filament which has the property of nearly constant current flow for widely varying voltages across it, and it is used on the Beckman type 72 pH meter and the E.I.L. type 23A pH meter. The Pye type 79 pH meter uses a different method, having the valve heater supply controlled by a regulating transistor and two zener diodes. Another feature of this and other instruments is the taut band suspension system used in the indicating meter, which results in a pivotless but robust meter of high sensitivity.

**Temperature compensation**

Provision is also made for automatic temperature compensation using a resistance thermometer. This is immersed in the solution under test, the resistance of the thermometers winding changing with temperature and so automatically varying the meter sensitivity. The adjustment of buffer pH in this pye circuit is different to the previous method as the buffer control gives a potential that is applied to the input in series with the EMF produced by the electrode cell.

The resistance thermometer is widely used for temperature compensation and is usually included in the feedback loop of the amplifier so that its gain is varied by the thermometer resistance changing with temperature.

The E.I.L. type 23A pH meter shown in figure 2.2(b) consists basically of two separate but identical d.c. amplifiers and though valve operated is still widely used. It should be noted that in this instrument the resistance thermometer has two separate windings, one being used to compensate for the change of slope of the pH-EMF curve by modifying the gain of the amplifier, the other being connected to a resistance bridge the balance of which is changed by the resistance thermometer to automatically compensate for the shift of the zero of the pH-EMF curve.

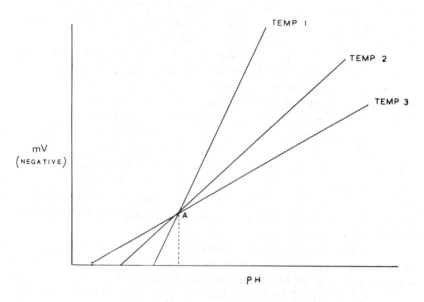

Fig. 2.2. (a). Variation of pH-EMF curve with temperature.

The input from the glass electrode is applied to the electrometer valve V1 and the temperature correction signal from the bridge network and resistance thermometer is fed to the grid of V3. If the temperature of the solution rises, the resistance of the thermometer increases and a positive temperature dependent bias signal is applied to V3 grid from the bridge network. The positive bias signal acts to oppose the higher asymmetry potential applied from the glass electrode to the grid of V1, thus providing automatic temperature compensation.

The gain of the amplifier is also varied by portion Ra of the resistance thermometer which is in the feedback path and provides compensation for sensitivity change. The isopotential control R1 enables the instrument to be calibrated against a buffer solution at one temperature and used at any other temperature within approximately 20°C by correctly setting the controls.

For additional stability the currents through the electrometer V1 and dummy input valve V3 heaters are maintained at a constant value by means of barretter V5 although the bulk of the stabilisation is performed by the constant voltage transformer CVT1.

A constant voltage transformer core has three limbs, the primary winding being on the centre limb. Main and subsidary secondary windings, connected in series opposition, are wound on the outer limbs. The magnetic circuit of the subsidiary winding is broken by a small air gap, so that over the working range of the input voltage, this gapped core is not saturate '. The magnetic circuit of the main winding is saturated and the output voltage may be held to within 1% for ± 15% change in the input voltage. The output voltage waveform is however non sinusoidal and only approximates to a sine wave if correcting filter circuits are used.

## Automatic standardisation technique

Standardisation and stabilisation for zero drift are done automatically with a special amplifier in the Beckman zeromatic pH meter. The electronic circuit is shown simplified and in the automatic operation mode in figure 2.3, and can be considered as two separate parts, (1) the measuring amplifier and (2) the corrector amplifier.

The measuring circuit comprises electrometer input valve V1, and together with valve V2 forms a three stage d.c. amplifier. The corrector amplifier is operational only when the relay RLA is closed and comprises V1, V2a, V3. Feedback maintains the input grid of V1 at a constant potential while capacitor C1 is charged. The meter reading indicates a voltage which is the sum of the signal volts Es + volts Em on memory capacitor C1 + any error voltage Egg; thus by charging capacitor C1 up to a voltage of Egg by feedback during the corrector cycle, we make Em equal Egg. On the measuring cycle the voltage on memory capacitor C1 therefore balances the error voltage Egg, so that only the signal voltage from the electrodes is indicated by the meter.

The relay which switches between 'measure' and 'correct' once a second, causes the corrector amplifier to be in circuit for about 0.015 sec, the relay being pulsed from a gas trigger tube.

A simple filter circuit in the output stage of the measuring amplifier removes the correction pulses so that they do not appear on the meter. In the measuring circuit, a signal at the amplifier input causes a voltage change at the cathode of the output valve and current flows through the meter and feedback resistance to ground, the feedback voltage being applied to the reference electrode. When equal to the voltage developed between the glass and reference electrodes, this voltage maintains a steady current in the output circuit. For temperature compensation the total feedback resistance is changed by the change of resistance with temperature of resistance thermometer TH.

## Synchronous chopper instruments

The use of direct coupled amplifiers has two main disadvantages,

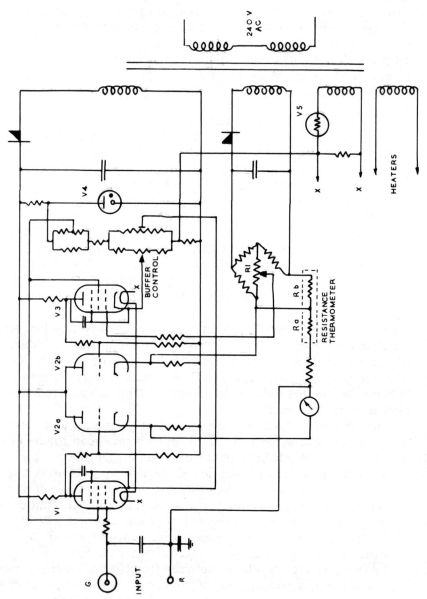

*Fig. 2.2. (b).* Simplified circuit of E.I.L. type 23A pH meter.

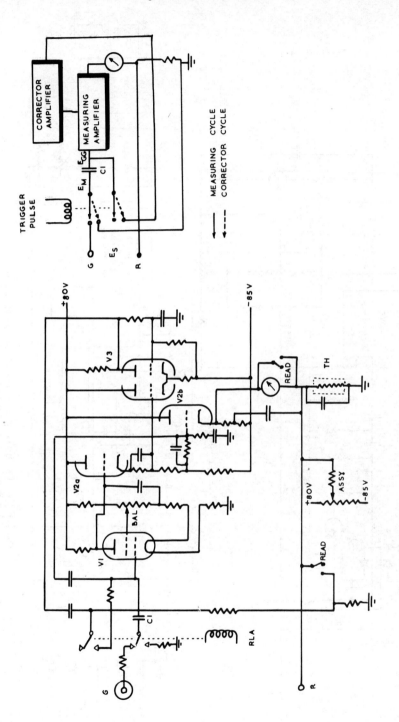

*Fig. 2.3.* Simplified circuit of Beckman Zeromatic pH meter.

38 ELECTROCHEMICAL INSTRUMENTS

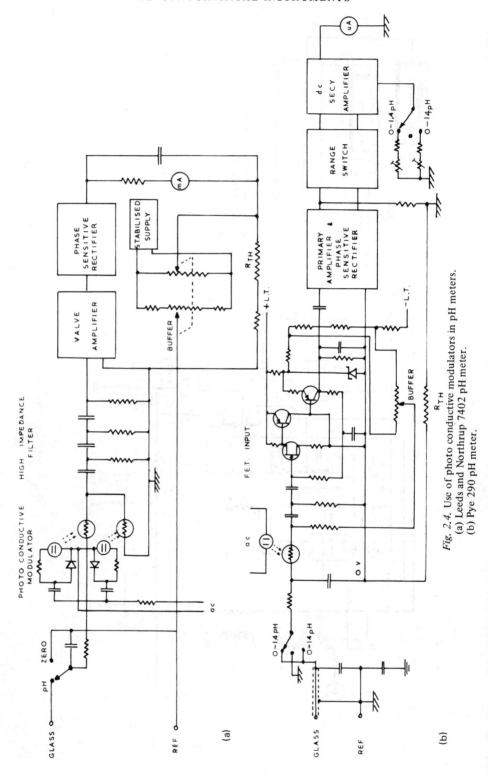

*Fig. 2.4.* Use of photo conductive modulators in pH meters.
(a) Leeds and Northrup 7402 pH meter.
(b) Pye 290 pH meter.

(1) zero instability and (2) electrometer valve grid current, which limit the stability of a pH meter to about 0.02 pH per day after allowing for valve warm up time and using stabilised power supplies.

It is possible to apply the technique of converting the d.c. signal to a.c. for stable amplification, as with other types of instrument. This may be done by using a synchronous chopper or a vibrating capacitor the input resistance of which is limited only by its d.c. insulation resistance and may be greater than $10^{13}$ ohms.

Many pH meters employ a vibrator or chopper to convert the d.c. signal to an alternating form. The chopper is usually a contact modulator, in which an electro-magnetically controlled oscillating contact, alternately makes connection with two fixed contacts.

This type of modulator gives rise to special design problems as the rapid making and breaking of the glass electrode connection gives rise to a series of voltage surges resulting in spikes being superimposed on the wave form. These spikes increase as the contacts age and their surface condition changes. All of these factors must be minimised in order to obtain trouble free operation and care is needed in the construction and choice of contact material of mechanical choppers.

Some manufacturers have recently introduced a solid state photo conductive modulator with no moving parts which overcomes these problems. The problem of the making and breaking of a direct connection between the glass electrode and amplifier, is also overcome in the vibrating capacitor circuit. The vibrating capacitor electrometer has a lower drift rate due to variation in contact potential with time than the mechanical chopper. In order to maintain this however, the capacitor plates must be chemically identical and sealed in a vacuum or dry inert atmosphere. They are therefore expensive.

Many types of pH meter use contact modulators successfully among them being Radiometer, Beckman and Metrohm. The Beckman research pH meter which uses a synchronous chopper of the contact type is shown in figure 2.6 and the circuit is described under the heading of potentiometric instruments.

The basic photo conductive modulator consists of a compact unit containing a neon lamp circuit which illuminates a photo conductive cell. A rectifier is connected in series with the lamp and when a.c. is supplied to this circuit, the lamp is extinguished during one half cycle of the supply. The lamp illumination thus varies in sympathy with the mains supply and the photo conductive cell oscillates between its high resistance (dark) state and low resistance (illuminated) state.

In order to replace the S.P.D.T. electro mechanical chopper two lamp circuits and photo conductive cells are necessary, figure 2.4(a). The rectifiers are connected so that one lamp is illuminated and the other extinguished on positive half cycles of the a.c. input. During

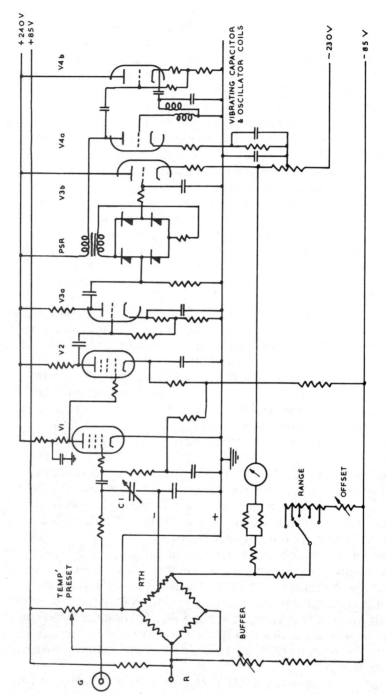

*Fig. 2.5.* Simplified circuit of the Pye Dynacap pH meter.

the negative half cycle conditions are reversed. The photo conductive cells are connected so that on one half cycle of the a.c. input the glass electrode is connected through a low resistance to a capacitance resistance filter. On the next half cycle this photo conductive cell reverts to its high resistance state and the other cell (connected from the capacitance input of the filter to the amplifier common line) is illuminated and in its low resistance condition. Circuits of the Leeds and Northrup pH meters and the Pye 290 pH meter, both of which use photo conductive modulators are shown in figure 2.4.

**Vibrating capacitors**

Using a vibrating capacitor the E.I.L. Vibron pH meter has a specification giving a zero stability of 0.005 pH over 24 hours after a 30 minute warm up period. Due to an a.c. amplifier being used the effect of mains voltage and frequency are negligible. Figures quoted for the Pye Dynacap which also uses a vibrating capacitor illustrate this very well, a mains voltage change of ± 15% giving a resultant pH shift of only 0.005 pH units while a change in the frequency of 2 cps gives a shift of less than 0.02 pH.

A typical vibrating capacitor consists basically of a metal reed oscillating near a fixed plate. The oscillation is sinusoidal, the drive coil operating from a low voltage 50 Hz supply or an electronic oscillator. The two capacitor electrode surfaces are separated by a small air gap, the surfaces being polished and gold plated.

The circuit of the Pye Dynacap vibrating capacitor pH meter is shown in figure 2.5. The plates of the vibrating capacitor are represented by C1 and the resulting a.c. signal is amplified by valves V1, V2 and V3a before being applied to the phase sensitive rectifier PSR.

Valve V4 is an oscillator which provides power for driving the vibrating capacitor coil and is also coupled to the transformer of the phase sensitive rectifier to provide a reference wave form. The rectifier output from the phase sensitive detector is applied to the output cathode follower valve V3b, which feeds the output meter.

Feedback from the output to the input of the a.c. amplifier is applied so that the effective input is reduced (negative feedback). As with the d.c. amplifiers already described, the gain of the amplifier is dependent on the resistance of the feedback loop, which includes a resistance thermometer for automatic temperature compensation.

Another instrument of this type is the E.I.L. Vibret model 46A pH meter. The output from the electrodes is converted to an alternating signal by the vibrating capacitor and then applied to the grid of an input valve. The electronic circuit which follows contains a four stage transistor amplifier and also a temperature compensation circuit.

Many pH meters incorporate scale expansion facilities and in one method using preset backing-off voltages, the meter zero can be set

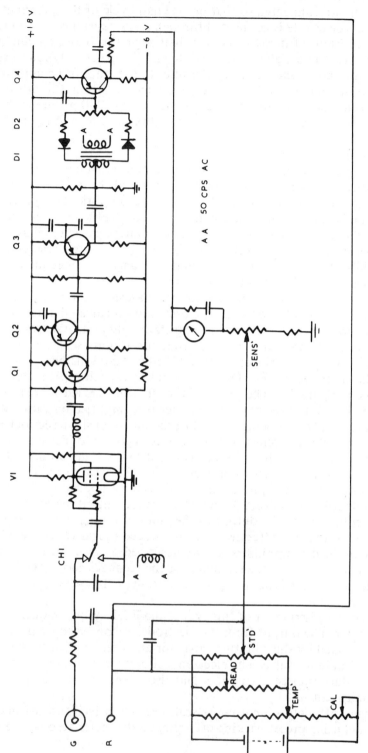

*Fig. 2.6.* Simplified circuit of Beckman research pH meter.

to any whole pH number over the range 0-12 pH. By this means and using a times ten sensitivity so that the full meter scale reads 1.4 pH, a reproducibility of 0.002 pH can be obtained.

The vibrating capacitor type of circuit gives excellent stability but for very accurate measurement of pH, potentiometric instruments are used which can measure to a relative accuracy of 0.001 pH units.

## Potentiometric pH measuring circuits

We shall consider two types, (a) a mains operated instrument using an a.c. amplifier and (b) a transportable battery operated instrument with a d.c. amplifier.

The Beckman research pH meter is a mains operated type and employs a slide wire potentiometer with an effective length of 12 ft. This potentiometer is mechanically linked with a readout dial which is 2 ft long for each pH unit of the instruments range of minus 0.5 to plus 14.5 pH. A null balance circuit is used (figure 2.6).

The d.c. electrode volts are passed through a filter and then converted to an alternating signal by a synchronous chopper CH1. An electrometer valve V1 converts the high impedance voltage signal to a current signal for the transistor amplifier which follows it. This amplifier comprises Q1, Q2, connected as a Darlington pair, and Q3 which feeds the phase sensitive rectifier transformer secondary winding. The transformer works in conjunction with diodes D1, D2 to convert the a.c. signal to d.c. which is then amplified by d.c. amplifier Q4 feeding the null indicating meter. The balancing potential is selected through a helical potentiometer which is directly linked with the readout scale. Manual temperature compensation is provided and the calibration potentiometer calibrates the slide wire against a standard cell.

The transportable Pye potentiometric pH meter is shown in figure 2.7, this instrument utilising a very accurate four decade potentiometer with manganin coils. This circuit also works on the null balance principle, and utilises an electrometer valve as the d.c. amplifier first stage, which is followed by transistors. Manual temperature compensation is provided over the range 0-100°C by means of a ten turn helical potentiometer, which has a discrimination better than 0.1 C.

## Field effect transistor instruments

It will be seen that valves are still used as amplifier first stages even in instruments utilising transistors although field effect transistors have now been developed and may be used as electrometers.

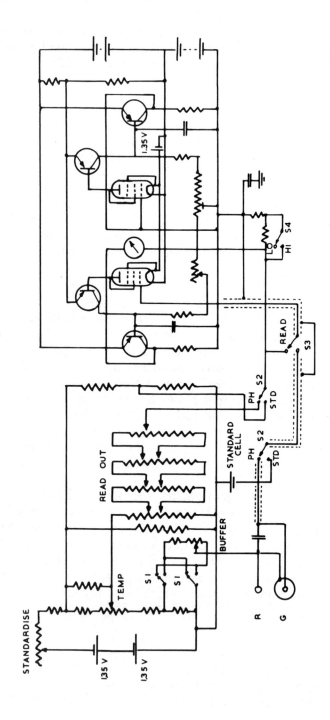

*Fig.* 2.7. Simplified circuit of Pye potentiometric pH meter (some switching omitted for clarity).

These devices are however expensive and at present there are few commercially available completely solid state pH meters.

The three element unipolar construction field effect transistor may have a gate leakage current of the order $10^{-8}$ amp which is much greater than the grid current of $10^{-14}$ amp obtainable with a vacuum electrometer valve, so that this type of field effect transistor is not suitable for electrometer work. The latest field effect transistor to be developed is the insulated gate type, which has an input impedance of about $10^{+12}$ ohms and a leakage current of the order of $10^{-12}$ to $10^{-15}$ amp.

It can therefore be seen that this type has characteristics comparable with the vacuum electrometer valve and is also mechanically more rugged. The insulated gate field effect transistor is however easily damaged by voltage surges and it is desirable to connect a permanent gate leakage impedance to prevent static electrical charges building up which could destroy the transistor.

A simple pH meter circuit using a field effect transistor in one arm of a Wheatstone bridge has been described by Jacobseen (55). The field effect transistor has a low drift rate with temperature and in this circuit without compensation gives a drift of about 0.05 pH per °C. With a thermistor connected to the drain, the drift is reduced to the order of 0.005 pH per °C. A 100 pF capacitor connected between the gate and source reduces a.c. pickup and the voltage surges which occur when the electrodes are removed from the solution. A commercial example of an all solid state pH meter is the Pye model 290, which uses a 2N3456 field effect transistor. A simplified circuit is shown as figure 2.4(b).

**Digital pH meters**

Digital voltmeter techniques in which the voltage input from the electrodes is converted to a form suitable for display on neon indicator tubes, have recently been employed in pH meters. Accuracy is typically ± 0.01 pH and the chance of reading errors are eliminated. A binary coded decimal output is available for feeding a printer or typewriter and the reset time between two measuring indications is about 1.5 seconds. This type of instrument should therefore be of value in clinical laboratories where a large number of readings may need to be taken and recorded every day. An instrument of this type is the transistorised Philips PW 9408 digital pH meter. Asymmetry potential (zero point) adjustment between 0.75 and 8.7 pH, manual or automatic temperature compensation and slope correction controls are provided and are used in the initial setting up. The screw adjusted slope correction potentiometer, calibrated 59.2 to 53.0 mV/$\Delta$pH (20°C), enables the amplifier sensitivity to be matched to the output of the electrodes. Performance figures for different types of pH measuring circuits are given in table 2.2.

| Type | pH ranges | Stability | Temperature compensation | Accuracy | Maximum Electrode resistance |
|---|---|---|---|---|---|
| *d.c. amplifier instruments* | | | | | |
| Pye 11067 | 0-8<br>6-14 | ±0.02 pH for mains voltage change of +15% | Manual & auto 0-100°C | ±0.1% | 1000 MΩ |
| E.I.L. 23A | 0-14 | ±0.02 pH over 24 hours | Manual & auto 0-100°C | ±0.05 pH | 500 MΩ |
| E.I.L. 38A | 0-8, 6-14 | ±0.05 pH over 24 hours | Manual & auto 0-100°C | ±0.1 pH | 1000 MΩ |
| Philips PR9401 | 1-9<br>2-12<br>4-14 | ±0.1 pH after 60 minutes | Manual | ±0.02 pH | |
| *Vibrating Chopper instruments* | | | | | |
| Radiometer PHM 28 | 0-10<br>6-14 | Negligible drift | Manual 0-100°C | ±0.02 pH (reproducibility) | |
| Metrohm E 322 | 0-14 in steps of 2 pH | Performance unaffected by ±15% mains voltage variations | Manual 0-70°C | ±0.01 pH | |
| *Vibrating Capacitor instruments* | | | | | |
| E.I.L. Vibron | 0-14<br>10 expanded ranges of 5 pH | ±0.005 pH over 7 days<br>Independent of ±10% mains voltage change | Manual & auto 0-100°C | ±0.005 pH (discrimination on most sensitive range) | 1000 MΩ |
| Pye Dynacap | 0-14<br>0-10<br>4-13<br>7 sub ranges of 2 pH | Less than 0.005 pH change for 15% mains voltage variation | Manual & auto 0-100°C | 0.02 pH (discrimination) | 1000 MΩ |
| *Photo Conductive Modulator instruments* | | | | | |
| Leeds & Northrup P 7403-A2 | 0-14 on any 2 pH span | Drift less than 0.005 pH on 2 pH span | Manual & auto 0-100°C | 0.005 pH in 2 pH span | 2000 MΩ |
| *Potentiometric instruments* | | | | | |
| Pye potentiometric | 0-13.999 in 0.001 steps | Requires standardisation only once a day | Manual 0-100°C | 0.001 pH | 1000 MΩ |

TABLE 2.2. *PERFORMANCE OF pH METER MEASURING CIRCUITS*

# POTENTIOMETRIC TITRATION

## Principles

The end point of a titration can be detected by many methods which may be grouped according to the property (EMF, current, conductivity, capacity, absorbance) measured.

In potentiometric titration, the EMF is measured of the cell formed by an indicator and reference electrode immersed in the titrated solution. The end point is reached when the rate of change of potential per unit addition of reagent is a maximum, or when a preset potential is attained.

In a neutralisation reaction, the change of pH is measured during titration using a glass electrode as the indicator. The end point may also be determined from the point of inflexion in the potential-volume curve. For oxidation-reduction reactions, an inert platinum electrode is usually used which acquires a potential dependant on the the logarithm of the ratio of the concentration of reduced and oxidized forms of the substance in solution.

Commercial potentiometric titration equipment consists basically of three parts, (1) a pH measuring unit (2) titration control unit and (3) burette/syringe delivery unit. Many different circuit techniques are used in the control units and some of these are outlined below. Facilities may also be provided for recording the titration curve.

Instruments may be divided into the following three groups
(1) Instruments in which the full titration curve of voltage as a function of volume is plotted.
(2) Instruments in which burette closure is at a preset end point potential.
(3) Titrators in which the end point is determined by a derivative of the voltage input signal.

With a control unit which is set to end the titration at a given potential, there is usually an anticipation circuit so that a change is made from fast to slow delivery of reagents as the end point is approached. In the pH-stat, which is of importance in biochemistry, a constant pH is automatically monitored and maintained by additions of acid or base.

## Recording instruments

In curve recording titrators, the titrant feed is usually from an electric motor screw driven syringe, which enables good synchronisation between flow and the recorder chart drive to be obtained.

An early design of this type was the Robinson potentiometric titrator which used a modified Leeds and Northrup recorder to control and operate a motor driven burette (94). The chart travel is made proportional to volume by using a synchro (selsyn) receiver and a synchro transmitter actuated by the burette motor.

Synchro units consist of a stator with three star connected windings (arranged similarly to those in a three phase motor), and a rotor carrying a single winding. With the rotor excited by an a.c. supply, a magnetic field is set up in the direction of the rotor axis. The field links with the stator windings and the EMF induced in them depends on the relative angle between the rotor and stator winding axis. The transmitter and receiver stator windings are connected together, and the current that flows, sets up a field in the receiver. If the transmitter rotor is moved, the receiver field rotates through an equal angle, and the receiver or follower rotor then aligns itself in the direction of the field repeated from the transmitter.

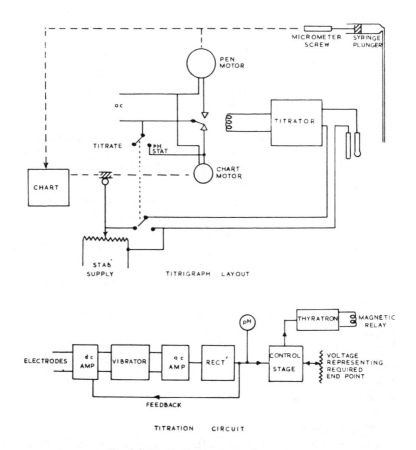

*Fig. 2.8.* The Radiometer Autotitrator.

This method forms the basis of Autotitrators developed by the Oak Ridge National Laboratory (109) (89). In the Q945 instrument, end point anticipation is provided by the amplified error signal from the electrode actuating a magnetic clutch, positioned between the burette electric motor and the synchro transmitter. The clutch thus controls both the titrant flow and the recorder chart whenever the error signal exceeds a certain value. A later instrument, the Q1728, uses a velocity servo system to control the velocity of the added titrant near the equivalence point (59).

These systems use a gating circuit which is opened and closed by a multivibrator to give an intermittent balancing feature. The error signal from the electrode passes through a chopper stabilised amplifier, and via the gating circuits operates the pen servo motor, which can only drive to balance while the gate is open.

There are of course many ways in which a recorder and syringe burette may be synchronised, and in the Radiometer Titrigraph a flexible shaft is used. A simplified diagram of this system is shown as figure 2.8, and it will be seen that two electric motors are used. The pen motor drives the recorder pen, and also by means of the flexible shaft, the micrometer screw mechanism of the glass syringe. The second synchronous motor drives the recorder chart and a control potentiometer, the output of which is thus proportional to chart movement. The titrator may be considered as a constant potential controller which makes the electrode potential follow the potential output of the control potentiometer.

The potential difference between the electrodes is applied to a vibrator circuit which is followed by an a.c. amplifier and rectifier. The d.c. output voltage is fed to a meter and to the control stage. In this stage, the amplifier output voltage is balanced against a potentiometer derived voltage, representing the required end point. The difference or error voltage is applied to a thyratron valve, the anode current of which actuates a magnetic relay. The relay contacts now de-energise the chart motor and energise the pen motor, so driving the syringe plunger and adding reagent.

In addition to the end point controls, a proportional band from the end point to a lower value may be selected. In this region the average flow of titrant is proportional to the error voltage, and the proportional characteristic is obtained by progressively reducing the on to off ratio of the magnetic relay as the end point is approached. The control stage is completed with a timing circuit which automatically shuts off this stage after a preselected number of seconds have elapsed without the addition of reagent to the sample. In addition to the syringe burette for titration recording, a magnetic valve is supplied to control the flow from any laboratory burette.

In the pH-stat mode, the control potentiometer is switched off. The electrode potential is constant as the titrant flow is controlled to compensate for any tendency of the process to change the pH of the sample. The chart motor runs continuously providing a time axis and the pen records the amount of titrant added.

An auto-burette unit giving a digital readout of delivered volume has been developed for use with the recorder and enables the rate of titrant consumption with time to be recorded.

A pH-stat with digital readout has also been described by Malmstadt and Piepmeier (76). This instrument uses a photoelectric sensor in the pen mechanism of an electrometric pH recorder, so that as the pH changes and the pen moves out of the light path, a relay is actuated. The relay unit allows reagent to be added, and stepping switches to operate an electromechanical counter, which counts the number of reagent aliquots added during a preset analysis time.

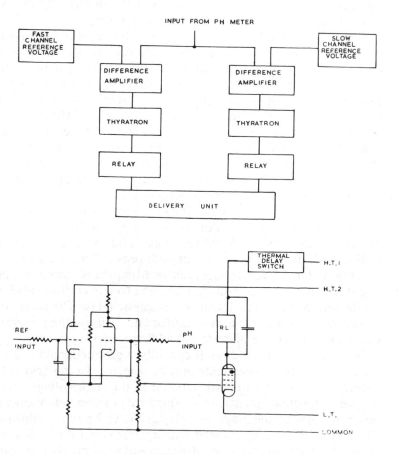

Simplified circuit of one channel.
*Fig. 2.9.* The Pye autotitrator.

The response of electrodes to adding an aliquot of reagent depends on the rate of stirring and the geometry of the electrode and burette/syringe tip. In the original Lingane titrator, end point approach was determined mainly by these factors, although in a later version the speed of the burette motor was automatically reduced by resistances being switched in to the motor circuit (71) (72). Transistorised and saturable reactor versions have also been developed (62) (63).

### Non recording burette closure titrators

Non recording titrators usually employ magnetic or solenoid

valves, which in the de-energised state, squeeze a length of flexible tubing so preventing the flow of titrant. Some form of anticipation control is incorporated, and in the Beckman model K titrator a resistance capacitance circuit is charged and then discharged on to the screen grid of an electrometer valve. The electrode potential is applied to the signal grid of the valve, the added potential on the screen producing a premature null response that via a relay temporarily closes the burette. On mixing, the electrode potential falls below the selected end point value and the charge/discharge cycle is repeated with the discharge time becoming longer and the charging time shorter, so that proper anticipatory end point approach is obtained.

Another method used by Pye Limited and shown in figure 2.9, utilises a delivery unit with two outputs, one giving a fast flow and the other slow. The titrator controller utilizes two difference amplifiers each of which feeds a thyratron valve in turn operating a relay. The relays control solenoids in the delivery unit and with no current flowing flexible tubes are squeezed so stopping reagent flow. Two independent potentiometer controls supply reference voltages for the difference amplifiers. One control (channel 1) is set to the point at which fast delivery is to stop and the other (channel 2) to the desired end point.

During titration the output from a pH meter unit is fed to the differential amplifiers and when the measured EMF balances the channel 1 potentiometer setting, the fast thyratron is fired and its relay operated so that fast delivery is stopped. Titration now proceeds at a slow rate until the end point, set on channel 2 potentiometer, is reached. At this point the slow thyratron fires and delivery ceases. Both channels of the instrument are independent, so that in order to control a reaction, one can be used to supply acid and the other an alkali.

**End point derivative titrators**

During titration the electrode EMF changes gradually except near the end point when a sharp change occurs. The end point may be located from the inflection point of the pH-volume curve, but more precise location can be achieved by plotting the rate of change of EMF against volume. In first derivative curves, the end point is at the maximum peak of the curve, and in second derivative curves at the zero point.

In the Philips PR9450 titrator the electrode signal was amplified in two stages, electrical resistance-capacitance differentiating circuits following each stage. The sudden drop and change of polarity of the second derivative voltage at the end point was used to actuate a valve operated relay system and close the burette.

Difficulties encountered with derivative titrators are the noise and distortion produced by successive amplification and differentiation. In the Philips circuit, which is now obsolete but of interest, the second electronic derivative voltage was applied to the grid of a valve with a relay in its anode circuit. During the positive portion of the curve the relay was energised, and its contacts used to charge a capacitor from a source of supply. As the voltage input passed rapidly through the zero point, and became negative going, the relay was de-energised and the capacitor discharged into a second relay. The contacts of this relay operated relay number 3, which actuated the burette solenoid. The second relay was only energised sufficiently if the capacitor had been fully charged. Spurious positive noise pulses, which are of short duration, at the valve grid did not then give false end points, as the first relay was not energised for sufficiently long to charge the capacitor.

Second derivative circuits have been described by Malmstadt, and a first derivative circuit by Shain and Huber (77) (95) (97). Instruments have been described using higher derivatives although it is doubtful if end point determination using derivatives higher than the third is worth while (75).

## AMPEROMETRIC TITRATION

**Electrodes**

The current flowing for a given applied EMF between an indicating and a reference electrode is measured in relation to the volume of added reagent, and L, reversed L, and V shaped titration curves may be obtained. The dropping mercury electrode (D.M.E.) or a rotating platinum electrode (usually motor driven at about 600 R.P.M.) are used as the indicator, with a mercury pool or calomel electrode as the reference. A good practical description of these and other types of indicator and reference electrodes has been given by Stock (103).

**Circuits**

In amperometric titrations there are two types of end point detection (a) 'kick off' and (b) 'dead stop'. In the first type, the current increases temporarily in steps as the end point is approached and at the end point there is a large current increase so that the measuring galvanometer kicks off scale. With the dead stop end point technique the electrode potential is fixed at a value which it will reach normally during the titration. The current decreases as titration proceeds and is zero at the end point due to the back E.M.F. from polarisation balancing the voltage applied between the electrodes. The most important application of the dead stop method is in the titration of water with Karl Fischer reagent.

The basic electrical equipment required for amperometric titration consists of a constant E.M.F. source of one or two volts and a galvanometer to measure the current. Amplifier circuits using a magic eye tuning indicator for end point detection have also been developed (81). Glastonbury has described a circuit using an audio frequency oscillator. The oscillator is effectively short circuited until the end point when an audible note is heard from the loud speaker (45). Transistorised circuits have also been described (90) (44) (104). A simple commercial automatic titrator is the Beckman KF-2 aquameter. The platinum electrodes form one arm of a bridge circuit and when the end point is approached the resistance between the electrodes drop. The bridge output thus decreases to zero then increases in the opposite phase as the end point is passed. The output is amplified and fed to a phase sensitive rectifier, the d.c. output of which operates a relay controlling the burette solenoid.

## COULOMETRIC ANALYSIS

### Principles

The measurement of current during a chemical process is the basis of coulometric analysis. In coulometric titration the reagent is not added from a burette but is generated by the electrolysis of a suitable solution. The end point may be detected by usual means, and the volume of reagent used up is calculated from the quantity (coulombs) of electricity required to generate it. This is simply done by using a constant current source and measuring the time of titration. Steady pulses may also be used instead of a steady current and the number of pulses counted (33).

Chemical coulometers or electronic integrating current amplifiers are other methods utilised. A low inertia integrating motor may also be used to determine the coulombs used. The motor is d.c. operated, with energy losses (heat, friction) reduced to a minimum, so that there is a linear relation between speed and voltage. If the applied voltage is derived from the voltage drop produced across a stable resistance, through which the electrolysis current is passing, and the motor shaft used to drive a mechanical counter, a direct reading, proportional to coulombs is obtained (11) (86).

Two basic methods are employed in coulometric analysis. These are (1) controlled potential and (2) constant current operation of the working electrode.

### Constant potential coulometry

In controlled potential coulometry, a coulometer, d.c. supply, a potentiostat and the electrolytic cell are required. The test substance may be oxidised or reduced at the working electrode (primary coulometry) or may react in solution with an electrolysis product (secondary coulometry). During this process the charge transfer is integrated by the coulometer circuit, which in modern electronic equipment is usually an operational amplifier connected as a capacitor integrator.

The potential of the working electrode is accurately controlled by the potentiostat, and as the reaction proceeds the current decreases from an initially high value to about zero on completion. The electrolysis cell consists of a working electrode and a counter or auxillary electrode to complete the circuit. A reference electrode positioned in the cell, is used to monitor the potential of the working electrode.

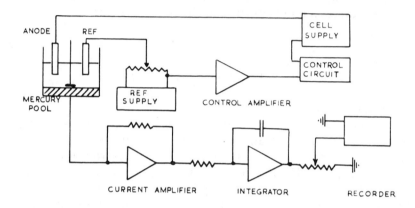

*Fig. 2.10.* Principle of electronic coulometric titrator.

The principle is shown in figure 2.10 of an operational amplifier circuit for controlled potential electrolysis and electronic integration. Many electronic coulometric titrators using these operational amplifier techniques have been described in the literature, the titration being complete when the current drops to a predetermined fraction of its initial value (79) (14) (57) (59).

The control amplifier senses the difference between the control potential and the potential of the controlled electrode, with respect to the reference electrode. The amplifier feedback loop is completed through the electrochemical cell, and the counter electrode is maintained at such a potential that the potential of the controlled electrode, with respect to the reference electrode, equals the control potential. The potential difference between the working electrode and counter electrode is not constant.

The circuit is shown for controlled cathode potential reductions, where the anode is a platinum wire electrode (isolated by a salt bridge and frit barrier) and the cathode a mercury pool. For controlled anode potential oxidations, the polarity of the control potential supply is reversed and platinum electrodes are used.

Electrolysis occurs at the counter electrode and products may interfere with titration. This is overcome by separating the electrodes in compartments, which are electrically connected by a salt bridge. Nitrogen is bubbled through the coulometric cell prior to use in order to remove oxygen.

The output of the current amplifier is proportional to the cell current and is integrated by the capacitor integrating amplifier. In the coulometric titrator circuit, the mercury pool is maintained at virtual earth potential.

**Constant current coulometry**

In the constant current method, only the elapsed time and current are required. Large surface area platinum electrodes are used in the titration vessel, which also contains electrodes for end point detection. The current may be monitored by measuring the voltage drop across a precision resistor.

Electronic constant current coulometric units, consist of a conventional rectifier and filter to give a d.c. voltage of a few hundred volts and a series regulating valve to control the current. The cell current develops a voltage drop across a standard resistance and is compared with the voltage from a constant reference source. Any difference is applied through an amplifier to the grid of the regulating valve. The current can usually be set to several constant levels by switching in other standard resistances. Electromechanical timers are also incorporated.

The end point may be detected by potentiometric, amperometric or photometric methods.

# CONVENTIONAL POLAROGRAPHY

**Principle**

In this method of analysis, a known and steadily increasing voltage is applied to an electrolytic cell, and the resulting current is measured. The current in simple apparatus is measured on a galvanometer, and does not increase continuously with the voltage but as a series of steps. The graph of voltage against current is known as a polarogram, and in order to obtain reproducible curves it is necessary that the electrode surface does not become covered by the products. This is achieved by using a dropping mercury electrode, or a development of it, so that the surface is self renewing. The mercury drops from the capillary tube at the rate of one drop every few seconds (65) (56) (22).

The current that flows initially is known as the residual current. The height of the wave or step is directly proportional to the concentration of the ions being reduced and the half wave potential is characteristic of the ion (figure 2.13). When a certain threshold potential is reached, ions are reduced and lose their charge to the cathode drop, giving rise to a current flow. The current increases rapidly with a further small increase in applied potential, due to the rate of reduction increasing. This continues until the reducible ions near the cathode drop are diminished, when the current levels out to a limited value, which depends on the diffusion of reducible ions. The limiting current is the sum of the diffusion current and a small migration current due to the potential difference between the anode and cathode. Adding a base solution having an excess of ions difficult to reduce, virtually illiminates the migration current.

As the amount of material removed from the solution is negligible, polarograms can be repeated many times without changing the solutions chemical composition. The current that flows is a few microamperes and it should be realised that the polarogram shape is determined by the processes occuring at the dropping mercury electrode. The anode is a mercury pool of comparatively large surface area, so that while the same current flows through it, the current density is very small. As the D.M.E. becomes more negative the potential of the anode does not change.

In the measuring circuit, the growth and fall of the mercury drop causes oscillations of the current which are damped out by using resistance capacitance networks. The residual current is significant with very low concentrations, and comprises a faradic current (due to the movement of ions under the influence of the electric field), and a condenser current (due to the mercury drop constituting a small condenser or capacitor). The latter current is only approximately proportional to the applied potential, while the faradic current depends on the presence of traces of oxygen or metallic impurities in the water used for making up the solution.

The faradic current can be reduced by care in preparation and the interference due to the condenser current, which makes the waves less well defined, can be eliminated by passing a counter current in the reverse direction. The residual current curve will then be parallel with the voltage axis. The diffusion current is dependent on the concentration of the solution and the characteristic of the D.M.E. These characteristics are the drop time and the mass of mercury flowing. This in turn depends on the pressure head, which should be maintained constant.

The temperature should also be constant as the wave height and the half wave potential change with temperature. The diffusion current may change 1½% per degree centigrade rise in temperature (111).

**Circuits**

The simplest polarograph consists of the cell, a potentiometric voltage supply and a damped sensitive galvanometer to measure the current (figure 2.11(a)). A photographic recording instrument

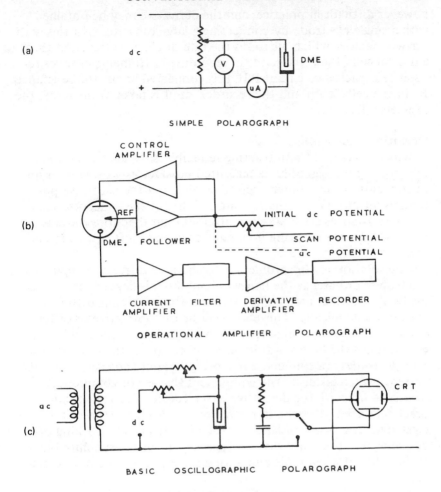

*Fig. 2.11.* Polarographic Methods.

(based on a design by Heyrovsky and Shikata) is marketed by E. H. Sargent & Co. and uses a motor driven potentiometer connected by gearing to a cylinder in a light tight housing. A photographic paper is mounted on the cylinder, the galvanometer movement carrying a mirror so that a light beam is focused on to the paper. The potentiometer and the cylinder rotate together, and a record of current against voltage is obtained as the galvanometer deflects (49).

Automatic pen recording instruments are also available, and variable RC damping, derivative facilities, and compensation for condenser (charging) current and interfering diffusion currents are usually provided (106). The derivative dI/dV of a polarogram can be obtained by using two matched D.M.E. electrodes, one of which is polarised more than the other. A galvanometer is used to measure the difference between the two electrode currents. This method is however difficult in practice, and the derivative may be obtained with a single electrode by connecting a large capacitor in series with a galvanometer, which measures the rate of change of current through a resistance in the cell circuit. In recording instruments, a series resistance capacitance circuit R1C1, is connected across the resistance Rc in the cell circuit and the recorder input is taken from across the resistance R1.

## Operational difficulties

With a mixture of non reacting reducible substances, the waves of a polarogram can only be measured satisfactorily using the type of conventional equipment described above, if the half wave potentials differ by more than about 150 mV. If the concentrations of the substances are markedly different (and therefore the wave heights) then the resulting polarogram becomes difficult to interpret.

One solution to this problem is to choose a different supporting electrolyte or base, as the half wave potential is dependent on this medium. Another, is to compensate for the interfering diffusion current electronically. This may be done by subtractive or differential polarography which require two D.M.E. electrode cells. One cell contains the base solution only and the other the test solution. The electrodes are connected in a bridge circuit and the differential current, which is due to the component that is not common, is measured. As with the derivative two electrode one cell circuit, practical difficulties arise in matching the characteristics of the electrode. The two capillaries must be identical and a striking off device is used to ensure that the mercury drops fall synchronously. If they do not, then node and antinodes appear on the polarogram (1).

Maxima, which resemble the overshoot trace obtained on a slightly under-damped recorder, sometimes occur on polarograms and may be supressed by adding surface acting agents such as gelatin, agar or dye ions (e.g. methyl red). The useful range of the D.M.E. is from $+ 0.4$ V to $- 1.8$ V, although this may be extended to $- 2.8$ V by using high reduction potential ions as supporting electrolytes. The charging or condenser current limits the range of conventional D.M.E. polarography to dilutions of $10^{-5}$ M. Derivative instruments are however about 100 times as sensitive.

Stationary solid micro electrodes may be used to overcome current oscillation, the platinum electrode covering the range $+ 0.9$ to $- 1$V. They have the disadvantage that the current flow is not constant. Rotating solid electrodes give a constant current but the wave plateaux may not be smooth. A hanging drop electrode in which the mercury drop adheres to the tip of a rotating wire projecting from a glass tube has also been developed (5).

The voltage applied to the polarographic cell is not the same as the potential difference between the electrodes. If E1 and E2 are the electrode potentials, then the applied voltage V equals E1 − E2 + iR, where i is the current and R the internal resistance of the cell. Correction must therefore be made for the iR voltage drop.

## CONTROLLED POTENTIAL POLAROGRAPHY

### Principle

Operational amplifier techniques have been applied to polarography and the circuit resembles that used for coulometric titration (37) (47). The cell contains a platinum electrode, calomel reference electrode and the D.M.E. (figure 2.11(b)). The control amplifier output maintains the potential of the polarised electrode (with respect to the reference electrode) equal to the control potential, regardless of variations in cell resistance. The control potential is derived from two supplies one giving the initial potential and the other a variable scan voltage. This may be derived from a motor driven potentiometer.

### Operational amplifier circuits

In the derivative controlled potential polarograph of Kelly, Jones and Fisher, the D.M.E. current is measured by a current amplifier which is followed by a diode capacitor circuit at the input to a voltage follower amplifier. In this way, the voltage follower output is the peak value of the input signal and current oscillations due to the growth and fall of the mercury drop are eliminated. The signal is next filtered, in order to average the polarographic current during the life of the drop, and then further amplified and differentiated before being fed to the recorder (60). The residual current is compensated for by the linear compensation circuit of Ilkovic and Semerano (54).

This method has the advantage of derivative polarography plus being independent of the cell resistance (3). In the conventional polarograms, good derivative curves are only obtained with low cell resistances up to about 1,000 ohms with aqueous solutions. With a high resistance, the trailing edge of the derivative curve is distorted and the peak height reduced (41) (73).

A circuit giving first, second, and third derivatives of the current from a hanging drop mercury electrode (H.M.D.E.) has been described by Perone and Mueller (88). The control circuit utilises two amplifiers and a sweep generator, which can give scan rates from 10 mv/sec to several hundred volts/sec.

The output from the H.D.M.E. is passed through a cathode bias amplifier to three differentiating amplifier units. Filtering is necessary because of the interelectrode noise pick up and noise generated in the differentiating circuits.

With stationary electrode polarography, the charging current is nearly linear with the scan rate, while the voltametric signal increases with the square root of the voltage scan. This gives rise to a bad signal to background ratio at high scan rates. This may be improved by using differentiating circuits and such a method has been described by Evins and Perone (40).

Methods of obtaining optimum performance with controlled potential polarographs using operational amplifiers, have been detailed by Booman and Holbrook. Optimum performance (which may be considered as the fastest possible rise time across the double layer mercury capacitance) with a given cell resistance and amplifier characteristics, can be obtained by the addition of passive networks and adjustment of the reference electrode position (15).

## a.c. POLAROGRAPHY

### Basic a.c. fundamental method

In the fundamental method of a.c. polarography, a small low frequency a.c. potential is superimposed on the d.c. supply to the electrodes. The a.c. components of the cell current can then be amplified rectified and recorded. A derivative type polarogram is produced, with peaks at the half wave potentials, proportional to concentration. The alternating current is small until the redox reaction occurs, when the current increases to a maximum, then falls when the limiting condition occurs. The peaks are clearly resolved even if the half wave potentials are close.

In a.c. polarography, background effects occur due to the capacitance of the cell and the double layer capacitance of the mercury drop (18).

The need to compensate for the charging currents and series iR drop, requires a knowledge of the resistive and capacitive components of the cell impedance, or of the phase angle and amplitude of the cell current. Phase angle measurements can be made with an impedance bridge, to determine the components of the cell impedance, or by using an oscilloscope to measure the phase angle and amplitude of the total cell current.

Hayes and Reilly (48) have described an a.c. polarograph in which the current and voltage are multiplied together, giving rise to a steady component, proportional to the in phase components of the current.

The multiplying voltage is the voltage across the cell minus the iR drop, so the steady component obtained is in phase with this voltage, which is not constant as it depends on the total cell current. If the phase of the voltage is moved 90° by a phase shift circuit, then the multiplication product is proportional to the quadrature component of the current.

The steady components of the multiplication process are not simply related to the in phase and quadrature components of the cell impedance, and in order to obtain proportional signals, the product of current and voltage is divided by a voltage proportional to the multiplying voltage. The resulting signal is again divided in a potentiometric recorder, the slide wire of which is fed with a d.c. voltage proportional to the multiplying voltage, so that recorder displacement is proportional to the admittance of the cell (reciprocal of impedance). A switch is provided in the circuit for quadrature and in phase measurements. The admittance, due to the double layer capacitance, is subtracted algebraically from the quadrature component of the total admittance. This method eliminates vector calculations necessary with the basic a.c. polarograph in order to allow for background effects.

Operational amplifier controlled potential circuits can also be used, and initial and scan d.c. voltages are summed with a small a.c. sine wave at the input of the control amplifier (99) (107) (110). The output from the polarised electrode is passed through a high pass filter to eliminate the d.c. component. The amplified a.c. current is then rectified, filtered and fed to the recorder. Operational amplifiers have a limited pass band and a further disadvantage is the introduction of noise by active circuit elements. This can be overcome by using a tuned amplifier which also enables other types of measurement such as phase angle and harmonic polarography to be undertaken. The introduction of a tuned amplifier brings other problems however, as if the sine wave oscillator frequency changes by only a small amount, a significant change in the output of the amplifier is obtained. The change depends on the Q of the tuned circuit and an oscillator stability of 1 in $10^4$ is desirable.

### Harmonic, phase selective and phase angle polarographs

Several second harmonic circuits have been described, and the usual method of obtaining the second harmonic current is by selective amplification of the current signal in a tuned amplifier. Smith has described some operational amplifier second harmonic phase selective and phase angle a.c. polarographs (100). In the phase selective second harmonic polarograph, the fundamental frequency output from the follower amplifier, connected to the reference electrode, is passed through a rectifier and tuned amplifiers to give a second harmonic reference voltage. This is applied to the reference input of a phase sensitive detector through a phase shift circuit. A second harmonic current signal is obtained by selective amplification in a tuned amplifier of the cell output current. This is applied to the signal input of the phase detector, the output of which is fed to a recorder.

In the operational amplifier a.c. phase angle recording circuit described by Smith, the output current and reference voltage sig-

62 ELECTROCHEMICAL INSTRUMENTS

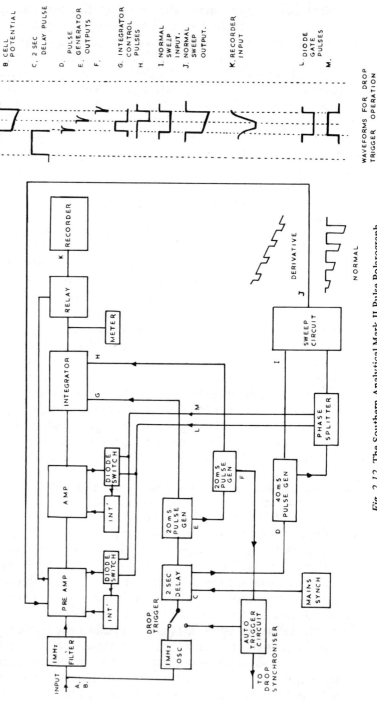

Fig. 2.12. The Southern Analytical Mark II Pulse Polarograph.

nals are passed through separate tuned amplifiers to trigger circuits. The two trigger circuits convert the signals to square waves whose amplitude is independent of that of the original sine wave. The square waves are added in a summing amplifier, and rectified in an analogue computer type absolute value rectifier circuit. This circuit's output d.c. component is dependent on the phase relationship between the square waves and is filtered and recorded. Paynter and Reinmuth have described 3rd and 4th harmonic methods (87).

Summarising the sinusoidal a.c. polarographic methods we have
(1) the basic a.c. fundamental current polarograph
(2) harmonic current component (2nd, 3rd or 4th) polarography
(3) measurement of phase angle between the fundamental current and applied potential.
(4) phase selective a.c. polarography in which the resistive in phase component of the fundamental harmonic current is measured.

A square wave instead of a sinusoidal wave may be superimposed on the d.c. polarising voltage. An advantage of the square wave is that the double layer capacitive current is less, in relation to the faradic current, than if a sine wave is used, thus offering easier separation. With square wave polarography reversibly reducible ions can be detected at concentrations as low as $10^{-6}$ M.

**Square wave polarographs**

Several square wave polarographs have been developed and Barker and Jenkins describe a circuit in which the current is measured just before each change in the square wave voltage (6). With a low cell resistance, the capacitive charging current has decayed to zero before the current measuring period, so that the capacitive current is eliminated from the polarogram.

A controlled potential operational amplifier instrument has been described by Buchanan and McCarten (23). The scan potential is obtained from an operational amplifier connected as an integrator, the output volts having a linear rise time. A phase shift oscillator is used to generate a sine wave, which is shaped, and then used to trigger changes in a multivibrator shift register circuit. The actual square wave is generated by a multivibrator which receives pulses from the shift register. The repetition frequency of the register and the square wave (100 Hz to 10 kHz) are controlled by the frequency of the oscillator. A gating circuit connected to the shift register allows a variable fraction of the measuring circuit output to be rectified and recorded.

**Pulse polarographs**

Sensitivities down to $10^{-8}$ M can be obtained for reversibly reducible ions using polarising pulses, each of which occurs at a definite time after the fall of the preceeding mercury drop. Pulse polarography was developed in order to obtain a higher sensitivity than that given by the square wave polarograph. The square wave method is limited, principally by instability in the response of the D.M.E., due to a thin film of electrolye tending to enter the capillary.

In the pulse polarograph originally developed by Barker, a pulse lasting $1/25$ second was applied to the electrodes 2 seconds after the fall of a drop which had a natural life of four seconds (7). The change in cell impedance at the drop fall was detected and used to initiate the timing cycle, the voltage pulse being applied during the stable part of the drop life. Subsequent development by Southern Analytical Limited, has reduced the delay period from 2 seconds to 1 second, with mechanical synchronisation of the drop with the polarograph. This is achieved by means of a solenoid which is operated at the end of the $1/25$ second (40 msec) pulse. The polarograph circuit is shown in figure 2.12.

For derivative polarograms a constant amplitude pulse is superimposed on a gradually increasing d.c. sweep voltage. For normal polarograms the cell d.c. voltage is constant, and the amplitude of the applied pulses is gradually increased from 0 to 1 volt.

The total cell current comprises a background diffusion current, the drop double layer capacity charging current, and the faradic current associated with the change of reaction rate as the pulse is applied. The background current is estimated instrumentally and subtracted from the cell current, and if the cell impedance is low enough, the charging current falls to zero in about 20 msec. Therefore, by measuring the current during the last 20 msec of the 40 msec applied voltage pulse, the polarogram becomes independent of the background and capacitive currents.

Several minor currents also occur which can affect the sensitivity. These are due to (1) changes in residual current, proportional to the charge supplied to the electrode when its potential is changed by the pulse, (2) a small charging current, associated with the mercury capillary, which occurs when the electrode potential changes abruptly, (3) oscillation of the drop producing modulation of the background current, (4) a current associated with organic molecules on the surface of the D.M.E. The origin and compensation for these currents is detailed in the literature (7) (83) (8).

With the Southern-Harwell mark II pulse polarograph, waves 40 mV apart can be resolved and reversibly reducible ions determined at concentrations less than $10^{-8}$ M. The sensitivity for irreversibly reducible ions is $5 \times 10^{-8}$ M. The current from the D.M.E. is amplified and then integrated, so that the ouput is a function of the current due to the last pulse or the last 3 or 9 pulses.

### The cathode ray polarograph

There are two main classes of instrument. In the first, the a.c. component of the current from an a.c. polarograph is examined.

oscillographically, and in the second a linear sweep voltage is applied to the D.M.E. and the cathode ray tube displays cell current voltage.

In the linear sweep circuit the potential developing across the cell is amplified and applied to the X deflection system of an oscilloscope (78). The cell current flows through a calibrated resistance and the resulting voltage is amplified and applied to the cathode ray tube Y plates. In the method suggested by Randles (92) and developed by Davis and Seaborn (28) (20) the start of the potential sweep is delayed until late in the life of the drop. This is for reproducibility of polarogram traces, as the rate of change of area of the drop is at a minimum when it is fully developed. A 7 second cycle is used, in which the cell voltage is held constant for the first 5 seconds, and then linearly increased by 0.5 volt. It is necessary to use a long persistance (afterglow) screen on the cathode ray tube and both normal and derivative traces can be obtained. With this method concentrations of $5 \times 10^{-8}$ M for both reversible and irreversible ions can be detected. With derivative curves the sensitivity is about $10^{-6}$ M.

The Southern Analytical A1670 cathode ray polarograph uses this technique. It has a deflection sensitivity $10^{-8}$ amps full scale, and after a 15 minute warming up time the drift of the electronic circuit is less than $5 \times 10^{-9}$ amp. Further development led to the design of the model A1660 differential cathode ray polarograph. In this instrument, two cells are used, both with automatically synchronised D.M.E.'s. Four modes of operation are possible. These are subtractive, comparative, derivative and single cell. In subtractive operation, one cell contains the sample, while the other cell contains either the supporting electrolyte or a suitable blank solution. For comparative work, one cell contains an accurately known standard of similar composition to the sample (30).

For derivative operation both cells contain the same solution and there is a small difference between the potentials applied throughout the voltage sweep. Waves 40 mV apart can be resolved in this way, and if a second derivative network in the amplifier system is used, a second derivative display is obtained enabling resolution of waves only 25 mV apart. By means of switching, the polarogram obtained may be due to either cell separately and can be of the normal or derivative form.

The cell current is converted to a.c. by an oscillator driven modulator, amplified and reconverted to d.c. by a phase sensitive rectifier. The signal is then passed through RC derivative circuits to the Y deflection system of the cathode ray tube. The sweep generator output is fed to the cell and the X deflection system of the cathode ray tube.

With these voltage sweep methods there is the problem of timing the sweep so that it always starts at the same age of each drop.

Snowden and Page have described a method in which the sharp change in current when the drop falls is used to actuate a time delay circuit (102). During the preset delay, the sweep volts are removed from the cell, and a blanking circuit is operated so that the spot is deflected off the screen of the cathode ray tube. This prevents a stationery spot burning the screen.

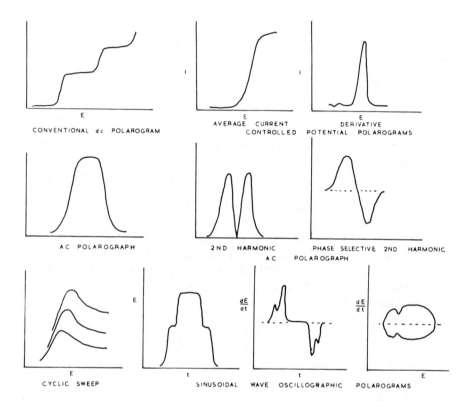

Fig. 2.13. Polarographic waveforms.

Another linear sweep method uses a saw tooth wave form which is applied with a quiescent period between sweeps. The horizontal deflection of the oscilloscope is made proportional to the voltage applied to the cell, and the potential developed across a resistance in series with the cell, is applied to the vertical deflection system. In the cyclic sweep method, several traces corresponding to various ages of successive drops are recorded. The waves are similar to ordinary polarograms and the largest peak, corresponding to the maximum drop area, is measured. The peak height is proportional to concentration (32).

Some of the first applications of the oscilloscope to polarography were described by Müller et al and Heyrovsky (84) (50) (51). If the amplitude of the a.c. volts superimposed on the d.c. voltage is made equal to the width of the conventional polarogram wave, and the d.c. voltage is exactly at the half wave potential, then a sinusoidal shaped trace is obtained. The alternating current component is applied to the vertical plates and on the horizontal plates the sweep frequency is synchronised with the applied voltage. The vertical height of the figure is a function of the diffusion current and therefore of concentration.

In the circuit of Heyrovsky, a series resistance and capacitance are connected to the D.M.E., and the vertical deflection amplifier is connected across them in order to obtain a polarographic peak trace Ef(t). The horizontal deflection system time base is synchronised with the voltage applied (figure 2.11(c)). With the oscilloscope connected across the resistance, derivative curves are obtained (dE/dt f(t)). If the cell voltage is applied to the horizontal deflection system an elliptical trace is obtained (dE/dt f(E)). Inflection indentations are made in the traces with a flow of current at the mercury electrode, and the curves are symmetrical for reversible reaction and non symmetrical for irreversible reactions. The oscillographic technique is a very sensitive method giving indication of relative rates of electrode reactions (82) (52). In order to avoid the problem of the changing area of the mercury drop a streaming mercury electrode is used (53). Some typical oscillographic polarogram traces are shown in figure 2.13. A comparison between square wave and oscillographic polarography has been made by Ferrett et al (42).

## POLAROGRAPHIC OXYGEN PROBES

**Principle**
The Clark type oxygen electrode is a fixed voltage (0.6 to 0.8 volt) polarographic probe for measurement of the partial pressure of oxygen in both gases and liquids. Oxygen diffuses through a thin (0.001 to 0.0005") teflon or polythene membrane, and is reduced at the platinum cathode, allowing a current to flow which is proportional to the oxygen contacting the cathode surface. The reference silver anode and the potassium chloride electrolyte (half saturated) are also behind the membrane so that no current flows through the solution being measured. A current of about one or two microamperes may be obtained from stirred air saturated water at $30°C$, with a polarising voltage of the order of 0.6 volt. If this current flows through a suitable resistance, sufficient voltage is developed to be fed into a 1 mV potentiometric recorder (31) (25).

**Operating techniques**
In use, a small amount of oxygen is consumed by the probe and as diffusion of oxygen in liquids is too slow to replace the con-

sumed oxygen, it is necessary to move the liquid past the probe tip by using a laboratory stirrer. For satisfactory operation the membrane must be smooth and free from wrinkles, and no air bubbles should be allowed to become trapped. It is usual to add one or two drops of Kodak Photo-Flo solution to the potassium chloride in order to ensure wetting. If the silver anode becomes dull it may be brightened using a $NH_4OH$ solution for up to one minute, followed by brushing with a cotton swab, and rinsing with distilled water. The electrode response is a function of the membrane and to check the response time, add sodium dithionite crystals to the solution. If the response is sluggish, change the membrane. A faulty membrane will also give rise to a noisy trace on the recorder. Normally membranes require frequent changing, but in a cuvette with a side mounted electrode, described by Sugarman and Appleman, a long membrane life is quoted (98). Oxygen probes have a temperature coefficient (membrane characteristic) of about 4% per °C, and to overcome this a thermistor may be built into probes used with portable amplifier units, such as the Beckman 777 and Yellow Springs Instrument Company model 51 Oxygen meter.

### Mackereth oxygen probe

A different type of probe (available from E.I.L. Ltd and developed by the Freshwater Biological Association and Water Pollution Research Laboratory) for use in lakes and rivers is called the Mackereth oxygen electrode. It uses a polythene membrane and has a lead anode and large area silver cathode giving a current of about 220 uA for air saturated water at 20°C. The probe has a temperature coefficient of 6% for each degree centigrade change and a separate thermistor probe is used to provide temperature compensation. The thermistor is included in the feedback path of the oxygen measuring amplifier and varies the amplifier gain as the temperature of the water varies. The circuit provides compensation within the limits − 5 to + 30°C for temperature variations of ± 10°C from a nominal value.

### Microelectrodes

Extremely small oxygen electrodes are also available and one example is the Beckman oxygen microelectrode, used for in vivo measurements of blood or body fluid dissolved oxygen. It consists of a 0.0005" platinum wire, sealed in glass, with its tip exposed and cemented in a silver tube. A polyethylene membrane is used to cover the platinum tip, and the whole assembly will fit into an 18 gauge arterial needle. Stirring is not necessary with microprobes as consumption of oxygen is negligible. Larger electrodes can of course be used for in vitro measurements and in the Beckman Macro electrode a 0.001 polypropylene or 0.00025 teflon membrane is used.

# CONDUCTIVITY MEASUREMENT

## Principle

To determine the conductivity of a solution its electrical resistance is measured and since direct current produces changes in the electrolyte, current pulses or a.c. must be used in the measuring circuit.

A form of the Wheatstone bridge circuit is usually used and a low frequency (50 to 5,000 Hz) a.c. voltage applied. The conductivity cell forms one arm of the bridge, the other three arms being resistive, although one arm utilises a variable capacitance in parallel with the resistance to balance out the capacitance of the cell. This calibrated resistive arm is variable and is adjusted until the bridge is balanced and no output obtained. The balance point may be detected by devices such as an a.c. galvanometer, earphones or a magic eye indicator.

Mains voltage frequency may be applied to the bridge through a step down transformer, or an oscillator used for higher frequencies, which are desirable when measuring low resistances in order to avoid polarisation or electrolysis.

## Instrument circuits

In some instruments (e.g. LKB conductolyzer, Philips conductance meter), the bridge output is amplified and applied to a phase sensitive rectifier which gives a d.c. output for operating a magic eye tuning indicator. This method eliminates errors not in phase with the oscillator signal, and discriminates between positive and negative changes in the resistance. Instrument controls may be calibrated in ohmic units or siemens, and a typical measuring range is from 0.1 uS to 10 S (10 M$\Omega$ to 0.1 $\Omega$).

Miniature cathode ray tubes may be used for null detection and an example is the Pye conductance bridge, in which the tube trace is in the form of an ellipse (in the horizontal position for balance). Errors in phase balance, produce a broader ellipse and for very accurate measurement a phase balance control is provided.

Conductivity changes with temperature, and temperature compensation circuits have been described by Dobos and Turicsin, which use temperature compensation resistors immersed in the conductivity cells (35). Another approach is to use a temperature sensitive element in the bridge circuit whose resistance changes with temperature at the same rate as the solution.

Platinum electrodes are used in the conductivity cells of which there are many types. The platinum electrodes are lightly coated with platinum black to reduce polarising effects, and for solutions of low conductance the surface area should be large and their spacing close. The reverse is the case for solutions with high conductivity.

A high sensitivity recording conductometric titrator with automatic temperature compensation has been described by Mueller et al using a transistorised LC oscillator operating at 1,000 cps (85).

## HIGH FREQUENCY TITRATORS

**Principle**

In high frequency methods of analysis, oscillator frequencies greater than 1 MHz are used, due to the relationship between the electrolyte concentration and the frequency for maximum sensitivity. At these frequencies, changes in the composition of the sample affect the electrical characteristics (conductance and capacitance) of the cell, which in turn affect an oscillator circuit.

Several types of circuit may be used, and at the lower frequencies where the response is determined by the sample conductance, the cell may actually form part of a tuned circuit (24). When operating at several megacycles or above, response is determined mainly by capacitance changes in the cell causing a change in the resonance frequency of the oscillator (93) (36). The change in oscillator frequency may be measured directly, using a frequency discriminator giving a voltage output proportional to the frequency change. Another method, mixes the outputs from a fixed frequency ($f_R$) oscillator with the output from an oscillator whose frequency ($f_O$) is varied by the capacitance changes in the cell. The mixer stage output is a beat or difference frequency ($f_R - f_O$) which may be measured on a frequency meter or on an oscilloscope. In all these methods, precision variable and switched capacitances, in parallel with the cell circuit, may be used to compensate for the change caused by the solution, so giving a zero or null reading at the output.

**Instrument circuits**

Many high frequency oscillator circuits are described in the literature, and in many instruments the change of anode or grid voltages are measured by a vacuum tube meter (46) (26). An example of this type is the Radelkis OK302 oscillatrator in which the test solution is between the plates of capacitive measuring electrodes. The resistance change of the solution affects the Q of the oscillator circuit (operating at 150 MHz), and changes in oscillator voltage are fed to a vacuum tube volt meter.

The Sargent oscillometer is an instrument of the null balance type, in which the capacitive and resistive contribution made to the circuit by the sample, are measured as the equivalent increments of a calibrated capacitance required to compensate for these effects. The circuit is shown in figure 2.14. At balance, the cell capacitance and the paralleled calibration capacitance act additively to deter-

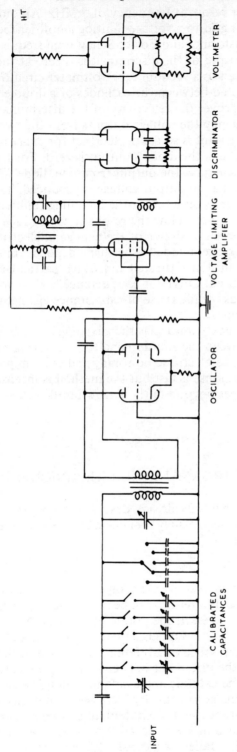

*Fig. 2.14.* Simplified circuit of the Sargent Oscillometer.

mine the oscillator resonant frequency of 5 MHz. An output is taken from the oscillator grid to a voltage limiting amplifier feeding a double tuned transformer and Foster Selley type frequency discriminator. The output of the discriminator is zero at the resonant frequency and is fed to a vacuum tube voltmeter circuit with a centre zero meter connected between the cathodes of a double triode valve. Double tuning improves the selectivity of the discriminator, and at the resonant frequency the voltage outputs from the two diodes are identical. The circuit is so connected that for exact resonance the difference between the diode's output is zero. For frequencies above or below resonance, the output from one diode will be greater than the other, so that an output voltage is produced. The polarity of this voltage depends on whether the incoming frequency to the discriminator is above or below the resonant value (12) (13).

Titration graphs can be drawn of the change in electrical response against volume of reagent added. The shape depends on the concentration of the substance titrated and which particular oscillator parameter is measured. Curves of conductance against volume generally have sharp peaks while those of capacitance against volume are of a broader V shape.

The high frequency titrator has the advantage that there is no direct contact between the solution and the measuring equipment so that effects due to electrode poisoning and contamination are eliminated. The response is fast but the method is insensitive to composition changes beyond fairly narrow limits.

## ELECTROANALYTICAL TECHNIQUES

In addition to the methods already described there are many other techniques used in electroanalytical chemistry. Some of these are briefly outlined below.

**Controlled potential**

For quantitative analysis by electrolysis with mixtures of metallic ions, a predetermined potential can be set so that one type of ion only is deposited. The electrode potential is controlled by a potentiostat. Many potentiostat circuits have been described and their operation requires the presence of a reference electrode in the cell. The potentiostat tests the difference between a reference voltage and the p.d. between the cathode and a calomel reference electrode, the applied E.M.F. being automatically increased or decreased in order to maintain a constant electrode potential. During electrolysis the current decreases as ions are removed and the separation is stopped when the current has fallen to a constant low value.

A solid state potentiostat for controlled potential electrolysis has been described by Lindstrom and Davies (70). The voltage is controlled to a preset value within the range $0 - 10$ V for currents up to 5 amps. Any tendency for the working cathode potential to change, causes an error signal to alter the charging of a capacitor in a unijunction transistor circuit. This in turn controls the point on the applied voltage wave form, at which a silicon controlled rectifier in the cell power supply circuits, fires. Booman and Holbrook have applied feedback theory and cell models to the design of controlled potential instruments (16).

Some other techniques are chronoamperometry (controlled potential technique with current against time recorded during electrolysis), rapid scan voltammetry (potential scanned linearly at a rapid rate the current against potential being recorded) and cyclic voltammetry (working electrode potential scanned linearly and periodically with time, the electrolysis current against potential being recorded).

### Controlled current

Another technique known as chronopotentiometry consists of electrolysing an unstirred solution at constant current, the potential of the working electrode against the reference electrode being recorded against time. The working electrode potential increases and is then stationary for a period known as the transition time followed by a further increase. The transition time is proportional to concentration.

## MULTIPURPOSE INSTRUMENTS

### Instrumental techniques

A multipurpose electro-chemical instrument for the control of potential or current has been described by Lauer et al (67). The potential between the reference and controlled electrode, is measured by differential electrometer amplifiers with high input impedance and common mode rejection. Current is measured by monitoring the p.d. across a precision resistor connected between the controlled electrode and ground.

Instruments have been developed which can perform a whole range of electroanalytical techniques (2) (4). Underkofler and Shain describe a multipurpose operational amplifier instrument for electroanalytical studies (108). Chopper stabilised operational amplifiers are used, except for the amplifiers which provide square wave and pulse outputs. The amplifiers can be connected for controlled potential analysis, polarography, constant potential electrolysis (low voltage power supplies are included for initial and pre-electrolysis

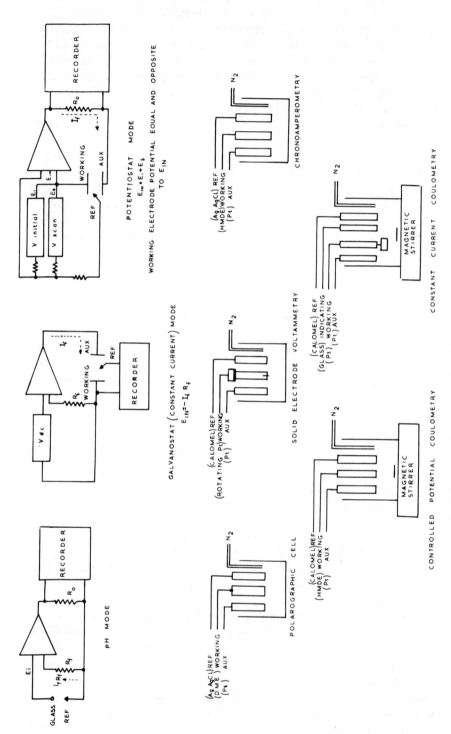

*Fig. 2.15.* Basic operational modes of the Beckman Electroscan 30.

potential in stripping analysis), a.c. polarography (sine wave of 20 – 100 Hz superimposed on linear scan), step function electrolysis, cyclic triangular wave voltammetry, coulometry, controlled current electrolysis and chronopotentiometry.

The recently introduced Beckman Electroscan 30 is a compact instrument combining measuring and recording facilities. Its basic operational modes are shown in figure 2.15 (58) (105). With the proper cell and electrodes and operated in the constant current mode; chronopotentiometry, coulometric titration, electro-deposition and stripping analysis can be performed. In the potentiostat mode; controlled potential coulometry, chronoamperometry, electro-deposition and separation, a.c. polarography, three electrode polarography, voltametry (solid electrode, rapid scan and cyclic) and stripping analysis techniques may be used. The recorder may also be used to measure pH and redox potentials.

The instrument contains a high gain operational amplifier circuit, and for controlled current operation, the voltage drop produced across a resistance in series with the cell is compared with a preselected potential. The difference is restored to zero through the operational amplifier circuit, the electrochemical cell being in the negative feedback path. The feedback potential at the working electrode (with respect to the reference electrode) is recorded. In the potentiostat mode, the potential between the working and reference electrode is compared with a preset potential. The amplifier and feedback circuit act to restore any difference to zero, and the voltage drop produced across precision resistors in the feedback circuit is recorded. There are five ranges of constant current, within the limits 0 – 100 mA, and the controlled potential supply provides ± 5 volts and a wide range of scan rates.

## Automation and computer techniques

Various techniques for automation of the acquisition and analysis of electrochemical data have been described. These involve conversion of the analogue input information to a digital form and the storage of the digital information in a memory store (usually magnetic). The stored information is extracted and converted into punched card or tape for analysis by a large computer (21) (68) (17) (19).

Lauer, Abel and Anson have described a system in which an on line digital computer is used as a combined controlled, acquisition and analysis device (69). Four operational amplifiers (potentiostat, current measurer, integrator, follower) are used in the data acquisition circuit, which is connected to the three electrode cell. An analogue signal is fed to an analogue to digital converter, and a clock circuit (accurate to 1 part in $10^7$) controls the gating and switching circuits, so that on completion of conversion, that value is transferred to the computer and stored. This technique was initially applied to double step chronocoulometry, in which a hanging drop mercury electrode is potentiostated at an initial value E and then abruptly stepped to a more cathodic value Ec. The resulting cathodic current is integrated, and after a time interval t, the potential is stepped back again to E and the resulting anodic current integrated. The computer is programmed to store 100 data points during both the anodic and cathodic current steps. The stored data is analysed by a least squares programme, and the slope and intercept of the least squares fitted lines are printed out on a typewriter.

Performance data of some typical electrochemical instruments are detailed in table 2.3.

| EXAMPLE | OPERATION | RANGES | PERFORMANCE |
|---|---|---|---|
| *Conductance Measuring Instruments* | | | |
| Philips PR 9500 | Bridge circuit and tuning indicator. Operating frequency 50 Hz or 1 kHz | 0.5-10 M$\Omega$ (6 ranges) | 2½% accuracy (0.2% for relative measurements) |
| Doran conductance meter | Bridge circuit and transistorised amplifier & phase sensitive detector. Operating frequency 1 kHz | 0-10 M$\Omega$ 0-10 $\mu$S (7 ranges) | 2% accuracy |
| Radiometer CDM 2 | Bridge circuit feeding amplifier. Operating frequency 3 kHz or 70 Hz | 0-500 $\mu$S 0-500 mS (5 ranges) | 1% accuracy |
| LKB conductolyzer | Bridge circuit feeding amplifier and phase sensitive detector. Operating frequency 2 kHz | 0-10 k$\Omega$ (3 ranges) | 0.25% accuracy |
| Pye 11700 | Amplifier and cathode ray tube for null indication. Operating frequency 300 Hz or 5 kHz | 0.1-10 M$\Omega$ 0.1 $\mu$s-10 S (4 ranges) | 0.2% accuracy |
| *High Frequency Titrators* | | | |
| Sargent Oscillometer | Oscillator feeding phase sensitive discriminator. Centre zero valve voltmeter. Operating frequency 5 MHz | 1000 division dial 5 range multipliers | |
| Touzart et Matignon High Frequency Titrimeter | Quartz crystal controlled oscillator. Q factor and anode current of oscillating circuit measured. Operating frequency 3 MHz | | |
| Radelkis OK-302 | Oscillator followed by vacuum tube voltmeter. Operating frequency 140 MHz | 6 step sensitivity control can detect 1 ml acid of $10^{-1}$ N dissolved in 100 ml water | |
| *Automatic Titrators (Potentiometric)* | | | |
| Radiometer TTT 1 Titrator | Used with pH meter & auto burette or magnetic valve. Flow rate reduced in proportional band (0-5 pH) | End point range 0-14 pH | 0.05 pH accuracy |
| Radiometer Titrigraph assy. | Used with TTT1 SBR2 recorder & SBU delivery unit | $\pm$ 0.5, 0.2, 0.1 pH/cm on 25 cm chart | 0.02 pH accuracy |
| Pye 11602 autotitrator | Used with pH meter. Differential amplifier and thyratron controls two outlet delivery unit. Delivery flow rate variable 2 ml/sec-0.1 ml/sec | End point range 0-14 pH, slow (anticipatory) delivery point range 0-14 pH | 0.1 pH (stability of control, point setting.) |

## MULTIPURPOSE INSTRUMENTS

| EXAMPLE | OPERATION | RANGES | PERFORMANCE |
|---|---|---|---|
| Metrohm Potentiograph E3B6 | Incorporates recorder and first derivative end point determination. Used with piston burette (delivery time – 3 to 80 mins.) | End point range 0-14 pH 0-10 0-5 | 0.2% accuracy |
| Sargent Recording Titrator | Incorporates electrometer amplifier and recorder. Used with motorised burettes. Delivery rates 10 ml/min to $1/15$ ml/min | 0-10 pH 0-5 0-2 | 0.25% accuracy |

### Dissolved Oxygen Analyzers

| EXAMPLE | OPERATION | RANGES | PERFORMANCE |
|---|---|---|---|
| Beckman 777 | Probe – silver anode, gold cathode Thermistor built in | 0-5 0-25 of $O_2$ 0-100 0-50 0-250 0-1000 mm$O_2$ partial pressure | 1% at constant temperature 5% from 15-45°C |
| Yellow Springs Model 52 | Probe – silver anode, gold cathode. Thermistor built in cell on instrument | 0-50% $O_2$ 0-380 mm$O_2$ | ½-1½% over full range |
| Yellow Springs Model 53 | Probe – silver anode, platinum cathode. No thermistor. Used with constant temperature water circulator | 0-100% AIR 0-100% $O_2$ | |
| E.I.L. Model 15A | Mackerith probe – silver cathode lead anode. Separate temperature compensator in probe holder. For use in lakes and rivers | 0-100% partial pressure of $O_2$ in water | 5% over 20°C range within limits –5 to +30°C |

### Polarographs (recording instruments)

| EXAMPLE | OPERATION | RANGES | PERFORMANCE |
|---|---|---|---|
| Metrohm Polarecord E 261 R | Combined recording and measuring instrument. Drop time control, can record dI/dt. Compensation circuit | Potentiometer voltage range –3 to +2 V. Starting voltage range –2 to +2 V 10 damping steps. Recorder ranges $2 \times 10^{-6}$ to $1 \times 10^{-10}$ amps/mm and 0-2000 mV | Can determine metal ions to $10^{-6}$ M |
| Sargent Polarograph XV | Combined recording and measuring instrument. Compensation circuit | Voltage ranges –4 to +1.5 V current ranges 0.003 to 1.0 μA/mm. 4 damping steps | Trace analysis of $10^{-6}$ M order with micro range extender |
| Radiometer Polariter PO4 | Combined recording and measuring instrument. Compensation circuit. Derivative facilities. | Voltage range 0-4 V current range $8 \times 10^{-11}$ to $8 \times 10^{-7}$ amps/mm 9 damping steps | |
| Southern Harwell A 1700 Pulse Polarograph Mark III | Combined recording and measuring instrument. Normal and derivative polarograms. Delay time, synchronisation, voltage sweep, polarising pulse height, measuring period and sensitivity variable | Starting voltage range +2.5 to –2.5 V Sweep voltage rate 1 V/min to 1 V/hour (6 steps). | Concentrations determined – reversibly, reducible ions $10^{-8}$ M, irreversibly reducible $5 \times 10^{-8}$ M |
| Southern Analytical A 1660 cathode ray Polarograph | Cathode ray tube presentation (long persistance screen). Normal, 1st and 2nd derivative curves. Camera attachment. Compensation circuit | Starting voltage range 0-2.5 V. Sweep voltage 0.5 V | Deflection sensitivity $10^{-8}$ A. Sensitivity (normal) $5 \times 10^{-8}$ M (derivative) $10^{-6}$ M |

TABLE 2.3.
*PERFORMANCE OF SOME TYPICAL ELECTROCHEMICAL INSTRUMENTS.*

## References

1. Airey and Smales (1950) *Analyst* **75**, 287.
2. Alden, J., Chamber, J. and Adams, R. (1963) *J. Electro. Anal. Chem.* **5**, 152.
3. Anniner, R. and Hugler, K. (1962) *Anal. Chem.* **34**, 362.
4. Bard, A. J. (1959) *Anal. Chim. Acta* **21**, 365.
5. Barendbrecht, E. (1958) *Nature* **181**, 764.
6. Barker, G. and Jenkins, I. (1952) *Analyst* **77**, 685.
7. Barker, G. C. and Gardner, A. W. (1960) *Zeitschrift für analylische Chemie* **173**, 79.
8. Barker, G. C. and Gardner, A. W. (1955) A.E.R.E. C/R 1606 H.M.S.O.
9. Beckman, A. O. The Development of pH Instrumentation. Beckman Instruments.
10. Beckman Instruments (1967) Specific Ion Electrodes.
11. Bett, N. (1954) *Analyst* **79**, 607.
12. Blaedel, W. and Malmstadt, H. (1950) *Anal. Chem.* **22**, 734.
13. Blaedel, W. and Malmstadt, H. (1953) *Anal. Chem.* **24**. 450.
14. Booman, G. (1957) *Anal. Chem.* **29**, 213.
15. Booman, G. and Holbrook, W. (1963) *Anal. Chem.* **35**, 12.
16. Booman, G. and Holbrook, W. (1963) *Anal. Chem.* **35**, 12.
17. Booman, G. (1966) *Anal. Chem.* **38**, 1141.
18. Brezer and Gutman (1945) *Aust. J. Sci.* **8**, 21.
19. Brieter, M. (1966) *J. Electrochem. Soc.* **113**, 1071.
20. British Standard 2586. H.M.S.O.
21. Brown, E., Smith, D. and DeFord, D. (1966) *Anal. Chem.* **38**, 1130.
22. Bryant and Reynolds (1953) *Analyst* **78**, 373.
23. Buchanan, E. and McCarten, J. (1965) *Anal. Chem.* **37**, 29.
24. Callan and Horrobin (1928) *J. Soc. Chem. Ind.* **47**, 329,
25. Clark, L. C. (1956) *Tran. Am. Soc, for Art. Int. Organs.* **2**, 41.
26. Clayton, J. C., Hazel, J. F., McNabb, W. N. and Schnable, G. L. (1955) *Anal. Chim. Acta* **13**, 487.
27. Conti, F. and Eisenham, G. (1965) *Biophysics J.* **5**, 247.
28. Davis, H. M. and Seaborn, J. E. (1953) *Electronic Engineering* **25**, 314.
29. Davis, H. M. and Seaborn, J. E. A.E.R.E. report R.3472.
30. Davis, H. M. and Rooney, R. C. (1962) *J. Pol. Soc.* **8**, 25.
31. Davies, P. W. (1962) The oxygen cathode in Physical Techniques in Biological Research. Vol. 4. Academic Press.
32. Delahay, P. (1950) *J. Phys. Chem.* **54**, 402.
33. Devanathan and Fernando (1956) *J. Sci. Inst.* **33**, 323.
34. Dobos, D. (1966) *Electronic Electrochemical Measuring Instruments*. Terra. Budapest.
35. Dobos, D. (1966) *Electronic Electrochemical Measuring Instruments*. Terra. Budapest.
36. Dowdal, J. P., Sinkinson, D. V. and Stretch, H. (1955) *Analyst* **80**, 491.
37. Durst, R. (1964) *Electro. Anal. Chem.* **7**, 245.
38. Eisenham, G., Mattock, G., Bates, R. and Friedman, S. M. (1966) *The Glass Electrode*. Interscience.
39. Eisenham, G. (1965) Advances in Analytical Chemistry and Instrumentation. C. N. Reilley. ed. Vol. 4. Wiley, New York.
40. Evins, C. and Perone, S. (1967) *Anal. Chem.* **39**, 309.
41. Ferrett, D. and Milner, G. (1955) *Analyst* **80**, 132.
42. Ferrett, D., Milner, G., Shalgosky and Slee (1956) *Analyst* **81**, 506.
43. Findlay, A. Practical Physical Chemistry. Longmans Green and Co.
44. Friedman, R. (1959) *Anal. Chem.* **31**, 1287.
45. Glastonbury, H. (1953) *Analyst* **78**, 682.
46. Hall, J. L. (1952) *Anal. Chem.* **24**, 1236.
47. Harper, K., Casimur, D. and Kinnersley, H. (1965) *J. Electro. Anal. Chem.* **9**, 477.
48. Hayes, J. and Reilley, C. (1965) *Anal. Chem.* **37**, 1323.

## MULTIPURPOSE INSTRUMENTS

49. Heyrovsky, J. and Shikata, M. (1925) *Rev. trav. chim.* **44**, 496.
50. Heyrovsky, J. and Forejt J. (1943) *Z. Physik. Chem.* **193**, 77.
51. Heyrovsky, J. (1953) *Anal. Chim. Acta* **8**, 283.
52. Heyrovsky, J. and Kalvoda, R. (1960) Oszillographische Polarographic mit Wechelstrom. Akademic. Berlin.
53. Heyrovsky, J. and Forejt, J. (1943) *Z. Physik. Chem.* **193**, 77.
54. Ilkovic, D. and Semerano, G. (1932) Collection of Czechoslov' Chem Commun'. **4**, 176.
55. Jacobseen, J. K. (1966) *Anal. Chem.* **38**, 1975.
56. Jolliffe and Morton (1954) *J. Pharm. Pharmacol.* **6**, 274.
57. Jones, H. C., Shults, W. D. and Dale, J. M. (1965) *Anal. Chem.* **37**, 681.
58. Joyce, R. J. An Introduction to Electroanalysis. Beckman Inst. Bull. 7079-A.
59. Kelley, M., Fisher, D. and Wagner, E. (1960) *Anal. Chem.* **32**, 61.
60. Kelley, M., Jones, H. and Fisher, D. (1959) *Anal. Chem.* **31**, 488.
61. Kelley, M., Jones, H. and Fisher, D. (1959) *Anal. Chem.* **31**, 1475.
62. Keyer, C. F. and Aronson, M. H. (1957) *Instr. and Automation* **30**, 1710.
63. Keyer, C. F. and Aronson, M. H. (1958) *Instr. and Automation* **31**, 474.
64. Kolthoff, I. M. pH and Electro Titrations. Chapman and Hall.
65. Kolthoff and Lingane. Polarography. Interscience.
66. Kuipers, P. Instruments and Measurements. Vol. 1. Academic Press.
67. Lauer, G., Schlien, H. and Osteryoung, R. (1963) *Anal. Chem.* **35**, 1789.
68. Lauer, G. and Osteryoung, R. (1966) *Anal. Chem.* **38**, 1137.
69. Lauer, G., Abel, R. and Anson, F. (1967) *Anal. Chem.* **39**, 765.
70. Lindstrom, F. and Davis, J. (1963) *Anal. Chem.* **35**.
71. Lingane, J. (1948) *Anal. Chem.* **20**, 285.
72. Lingane, J. (1949) *Anal. Chem.* **21**, 497.
73. Lingane, J. and Williams, R. (1952) *J. Am. Chem. Soc.* **74**, 790.
74. MacInnes, D. and Belcher, D. (1933) *Ind. Eng. Chem. Anal. Ed.* **5**, 199.
75. Maekowa, Y. (1956) *Pharm. Bull. (Tokyo)* **4**, 321.
76. Malmstadt, H. and Piepmeier, E. (1965) *Anal. Chem.* **37**, 34.
77. Malmstadt, H. and Fett, E. (1954) *Anal. Chem.* **26**, 1348.
78. Matheson and Nichols (1938) *Trans. Am. Electrochem. Soc.* **73**, 193.
79. Matsen, J. and Linford, H. (1962) *Anal. Chem.* **34**, 142.
80. Mattock, G. (1957) *Lab. Prac.* **6**, No. 8.
81. McCauley, C. and Gresham, W. (1955) *Anal. Chem.* **27**, 1847.
82. Meites, L. Polarographic Techniques. Interscience.
83. Milner, G. W. C. (1957) The Principles and Applications of Polarography. Longmans Green and Co.
84. Müller R. H. (1938) *Ind. Eng. Chem. Anal.* **10**, 339.
85. Mueller, T., Stelzner, R., Fisher, D. and Jones, H. (1965) *Anal. Chem.* **37**.
86. Parsons, J. S., Seaman, W. and Amick, R. M. (1955) *Anal. Chem.* **27**, 1754.
87. Paynter, J and Reinmuth, W. (1962) *Anal. Chem.* **34**, 1335.
88. Perone and Mueller. (1965) *Anal. Chem.* **37**, 2.
89. Phillips, J. P. Automatic Titrators. Academic Press.
90. Phillips, J. (1956) *Chemist-Analyst* **45**, 107.
91. Pungor, E., Havas, J. and Toth, K. (1965) *Instr. Control Systems* **38**, 105.
92. Randles, J. (1948) *Trans. Faraday Soc.* **44**, 334.
93. Reilley and McCurdy (1953) *Anal. Chem.* **25**, 86.
94. Robinson, H. A. (1947) *Trans. Electrochem. Soc.* **92**, 445.
95. Sargent, E. H. and Co. (1955) Scientific Apparatus and Methods Vol. 7. No. 2.
96. Severinghauss, J. W. and Bradley, A. F. (1958) *J. Appl. Physiol.* **13**, 515.
97. Shain and Huber. (1958) *Anal. Chem.* **30**, 1286.
98. Shugarman, P. and Appleman, D. (1967) *Anal. Biochem.* **18**, 193.
99. Smith, D. (1963) *Anal. Chem.* **35**, 610.
100. Smith, D. (1963) *Anal. Chem.* **35**, 1811.
101. Snell, F. M. (1960) *J. Appl. Physiol.* **15**, 729.
102. Snowden, F. C. and Page, H. T. (1950) *Anal. Chem.* **22**, 969.
103. Stock, J. (1964) Amperometric Titrations. Interscience.
104. Stock, J. (1958) *Analyst.* **83**, 56.
105. Sussman, A. (1967) Guide to Basic Applications of the Electroscan 30. Beckman Instruments.

106. Tinsley & Co. (1961) *Lab. Prac.* **10**, 537.
107. Underkofler, W., Shain, I. (1963) *Anal. Chem.* **35**, 1779.
108. Underkofler, W., Shain, I. (1963) *Anal. Chem.* **35**, 1779.
109. U.S. Atomic Energy Comm. ORNL Master Analytical Manual T1D-7015 (section 1).
110. Walker, D., Adams, R. and Alden, J. (1961) *Anal. Chem.* **33**, 308.
111. Willard, H., Merritt, L. and Dean, J. Polarography in Instrumental Methods of Analysis. D. Van Norstrand.

# 3 Spectrophotometric Instruments

### LIGHT SOURCES AND DETECTORS

The component parts of an absorption spectrophotometer consist essentially of a light source, an optical dispersing element to vary the wavelength of light passing through the sample, a photoelectric detector, and a means of amplifying the signal from the detector so as to operate a recorder or meter reading the percentage transmission or absorption (optical density) of the sample. It is necessary to stabilise the electrical supply to the amplifier and light source so that consistant reliable results are obtained.

**Wavelength response**

For work in the visible and ultra-violet range, tungsten and hydrogen (or deuterium) lamps are used respectively. In a typical quartz monochromator instrument, the tungsten lamp is used with a red sensitive photocell from 1,000 m$\mu$ to 625 m$\mu$, and with a blue sensitive photocell from 625 m$\mu$ to 320 m$\mu$. The hydrogen lamp, which has a quartz window, is used with the blue sensitive photocell from 185 to 320 m$\mu$.

Both the light source and detector tubes must be of a material, or have a window, which is transparent to the appropriate light wavelength. The blue sensitive photocell has a quartz window or is enclosed in a tube of fused silica. It is also necessary to ensure that the sample cell is of a material suitable for the wavelength used. Glass is satisfactory from 1,000 m$\mu$ to 360 m$\mu$, and standard silica down to 250 m$\mu$. Special silica or fused quartz cells are necessary for the far ultra-violet and infra-red regions.

**Tungsten lamps**

The tungsten filament lamp used for a light source in the visible region, is usually of low voltage and high current rating. The voltage is stabilised so that, for example, with a 15% change in the a.c. mains power supply, the low voltage d.c. supply to the lamp changes less than 1 part in 1,000. Berger et al. have described a transistorised voltage regulator with fast recovery for use with tungsten lamps (5).

The filament of the tungsten lamp is closely coiled in order to obtain the maximum brightness per unit area. Motor vehicle head-

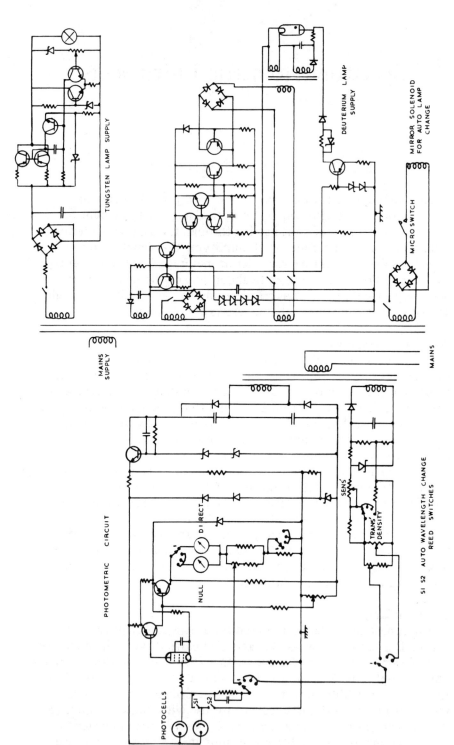

Fig. 3.1. The Unicam SP 500 Series 2 spectrophotometer.

lamp bulbs may be used and can be slightly overrun in order to increase the brilliance. Tungsten ribbon filaments are also available, but while they give brighter and more even illumination, they are expensive. Quartz iodine lamps of higher intensity may also be used. After a period of use, dark patches occur on the inside of the tungsten lamp bulb due to evaporation and deposition of tungsten. This of course, reduces the illumination, and together with pitting of the filament may give rise to erratic reading. The lamp should then be replaced.

**Stabilised lamp supplies.**

The stabilised lamp supplies used with the Unicam SP500 spectrophotometer are shown in figure 3.1. In this circuit, a portion of the voltage supply to the tungsten lamp, is compared against a highly stable reference voltage using a difference amplifier. Should the output change by a small amount, a difference exists between the two inputs to this amplifier and an output signal is produced which, after further amplification, controls the bias voltage on the base connection of the series controller transistor. Altering the bias on this transistor effectively varies its resistance, the circuit being so designed that should the lamp voltage change, then the resistance of the series controller varies in such a way that more or less volts are dropped across it in order that the lamp voltage is restored to its correct value. In practice several power transistors are connected in parallel so that the power dissipation of each transistor is not exceeded.

**Hydrogen and deuterium lamps**

The radiation output from a hydrogen lamp varies with the arc current, and it is important that the current remains constant in spite of variations in the arc voltage drop and line volts. A change in the lamp current causes a change in voltage which may be used via amplifier circuits, to change the effective resistance of the lamp supply circuit, so restoring the current to its correct value.

Hydrogen lamps used in spectrophotometers such as the Beckman DU, DB, DK2 and Unicam SP500, SP700 and SP800 are of the low voltage heated cathode type. These lamps, giving a continuous spectrum due to a discharge in hydrogen, have an anode, cathode and a filament heater, one end of the heater being connected to the cathode. It is usual to allow several minutes for the lamp filament to warm up before the arc between anode and cathode is struck. In the circuit of figure 3.1, both H.T. and filament supplies are applied directly to the lamp at switch on. These voltages are however removed by the diodes, in the supply line to the starting transformer, being changed to their high resistance state after a time period determined by a transistorised Miller circuit. The H.T. for the lamp is then supplied via a regulating transistor.

The hydrogen lamp is being replaced by the deuterium lamp which provides greater energy than a standard hydrogen lamp. Recent work by Lippman (37) and also by Levikov and Shishatskaya (36) shows that the ratio of light output from deuterium to that of hydrogen is about 1.5 over the range 2,500 to 3,000 Å (a 50% increase).

The life of a hydrogen lamp may be 200 hours, after which failure is indicated by a decrease in the light output, or by instability due to pressure changes caused by leakage. It should be noted that inability of the lamp to strike correctly is not always a sign of failure. This may be due to the lamp being too hot from previous use when turned on, so that ionisation occurs inside the anode, rather than in one concentrated spot in front of the window. The remedy is to turn the lamp off and allow it to cool.

For work in the far ultra-violet region most commercial spectrophotometers lamp housings can be sealed and purged with a non-absorbing gas.

### High intensity light sources

Mercury and xenon lamps may be used where high intensity illumination is required. The mercury lamp consists of a quartz envelope, containing two tungsten electrodes, between which an arc is formed in the high pressure mercury vapour. Energy is emitted in a number of discrete lines, and these lamps are used where a high degree of spectral purity is required. Table 3.1, below, shows the spectral lines emitted. The Eppendorf photometer utilises a mercury or a mercury-cadmium lamp, and filters are used to isolate the lines of interest. The filter is selected to be as near as possible to the wave length of maximum absorption of the sample. The output from the photocell is connected to an a.c. amplifier by a cathode follower stage, the amplifier output being converted to d.c. by a rectifier unit. The photometer unit output may be connected to a separate potentiometric recorder and a suitable circuit for this has been described by Estarbrook (20).

### Infra-red sources and detectors

Having described light source used in visible and ultra-violet spectrophotometers, we should mention that special techniques are necessary for infra-red work. Two of the common light sources in use are the Nernst glower and glowbar. The Nernst glower, consists of a rod of zirconium oxide with platinum wires embedded in the ends. This rod is normally non-conducting, but when heated to approximately 800°C, becomes conducting. The glowbar is a rod of silicon carbide, which conducts at normal ambient temperatures, so it can be heated by passing current through it. Nichrome wire is also used as an infra-red source.

*Principal Lines (mµ).* Emission from Beckman Mercury Lamp. Detection of lines above 600 mµ on spectrophotometer requires use of lead sulphide cell or red sensitive photocell.

| | | | |
|---|---|---|---|
| 184.968 | 313.269 | 407.893 | 1128.9 |
| 194.232 | 334.241 | 434.867 | 1357.37 |
| 253.726 | 365.115 | 546.220 | 1367.70 |
| 289.442 | 365.584 | 577.115 | 1395.43 |
| 296.811 | 366.428 | 579.221 | 1529.98 |
| 312.653 | 404.766 | 1014.24 | 1692.4 |
| | | | 1709.4 |

*Secondary Lines (mµ).* Emission from Beckman Mercury Lamp. Detection of lines requires high sensitivity settings on spectrophotometer and use of lead sulphide cell above 600 mµ. The higher energy peaks of the secondary lines are indicated by an asterisk.

| | | | |
|---|---|---|---|
| 177.3 | 229.701 | 248.453 | 284.848 |
| 179.95 | 230.277 | 249.276 | 285.775* |
| 182.073 | 231.507 | 253.550 | 289.442* |
| 186.924 | 232.388 | 256.459 | 291.709 |
| 197.389* | 234.124 | 257.703* | 292.623* |
| 198.793 | 234.612 | 260.390 | 302.234* |
| 202.709* | 235.317 | 262.599 | 302.431* |
| 205.356* | 237.901* | 264.003 | 302.645* |
| 214.865 | 238.069 | 265.280* | 302.833 |
| 219.140 | 240.005* | 265.444* | 370.518 |
| 222.537 | 240.120 | 265.589* | 380.270 |
| 224.836 | 240.805 | 267.567 | 390.743* |
| 225.345 | 241.480 | 269.960* | 410.917 |
| 225.943 | 244.174 | 275.356* | 434.040* |
| 226.073* | 244.761 | 276.049 | 435.950* |
| 226.290 | 246.477* | 280.427* | 491.739* |
| 226.431 | 248.271* | 280.524 | 502.699 |
| 227.059 | 248.342 | 280.756 | 567.742 |

*High Intensity Light Source.* Emission from H 100 A4 lamp. In addition to the above lines, the wavelength peaks listed below may be detected on a spectrophotometer with high sensitivity and using a lead sulphide cell.

| | | | |
|---|---|---|---|
| 690.930* | 1169.2 | 1213.2 | 1720.4 |
| 708.374 | 1177.2 | 1321.33* | 1733.4* |
| 773.054 | 1189.1* | 1343.0* | 1744.1 |
| 1071.8 | 1197.9 | 1350.9* | 1813.5* |
| 1118.0* | 1202.3 | 1707.7 | 1970.5* |
| 1149.5 | 1207.5* | 1711.4 | 2250.7* |
| | | | 2326.2* |

TABLE 3.1.

*MERCURY LAMP SPECTRAL LINES*

Infra-red radiation may be detected by (1) thermocouples or thermopiles (30), (2) photoelectric cells, (3) Bolometers (65) (24) (2), and (4) the pneumatic golay detector (25). In the golay detector, radiation passes through a window in a chamber containing xenon gas, which is warmed by absorbing the radiant energy. This causes a rise in the pressure of the gas, so distorting a flexible mirror forming part of an optical system, which in turn causes the intensity of light falling on a photoelectric detector to change. The bolometer consists of mixtures of manganese, nickel and cobalt oxides deposited on glass or silica. The resistance changes with temperature.

## Barrier layer photocells

The barrier layer photocell consists of a semiconductor layer deposited on a metal (e.g. selenium on iron). With a low external resistance (about 200 ohms or less) connected, there is a nearly linear relationship between the photo current and the incident illumination. The spectral response extends over the visible region. This type of photocell has several disadvantages; (1) low resistance, so that a cathode follower is necessary when amplifying the signal, (2) fatigue effect, (3) slow response to changes of illumination and (4) a high temperature coefficient. For these reasons it's use is restricted to simple filter photometers where a high level of illumination exists.

## Simple filter photometers

In the filter photometer, light from a tungsten source is passed through a filter, then through the cell on to a photoelectric detector. The indicating meter, fed from the photocell is set to 100% (with a blank in the light path) by using a diaphragm in the light beam, or by altering the lamp intensity with a variable resistance in the lamp circuit.

Two photocells connected in a bridge circuit may be used in a double beam arrangement, whereby the light from the source falls on the reference cell, and the light after passing through the sample on the measuring cell. A null balance galvanometer is connected across the bridge circuit. An instrument of this type is the Hilger-Spekker and a diaphragm is used between the light source and reference cell in order to set zero absorbance. A cam shaped diaphragm, between the source and sample cell, is connected to an external logarithmic absorbance scale, and is adjusted to compensate for the absorbance of the sample.

In these double beam arrangements, the lamp may be operated from the a.c. mains supply, as the circuit compensates for normal voltage variations. In the simple filter photometer or colorimeter constant voltage transformers are used for the lamp supply.

## Photo emissive cells

Photo emissive cells consist of either gas filled or vacuum tubes with an anode and a cathode, the latter having a special metal deposit. This is usually caesium antimony or a bismuth-oxygen-antimony-caesium mixture (51). Light striking the photo cathode causes electrons to be liberated; these are attracted to the anode which is main-

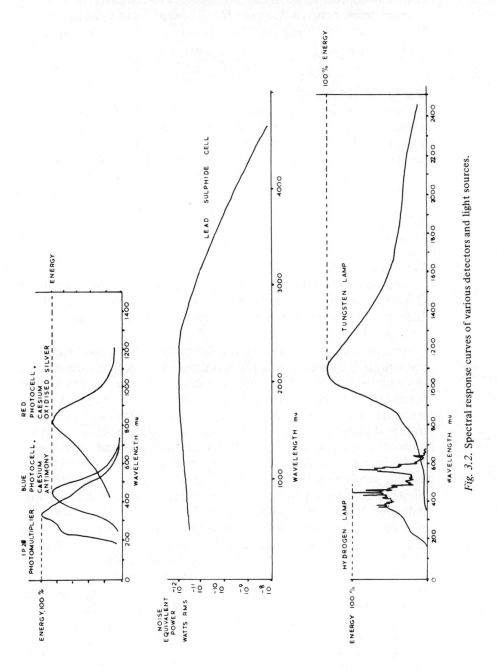

*Fig. 3.2.* Spectral response curves of various detectors and light sources.

tained at as positive potential with respect to the cathode. In spectrophotometry the vacuum photocell is used, as the photo current bears a linear relationship to the intensity of the light which falls on the cathode.

### Photoconductive cells

Photoconductive cells change their resistance when light falls on them. A typical example is the lead sulphide cell used on the Beckman DK2 spectrophotometer. This consists of lead sulphide, deposited on a quartz plate, and is used over the range 400 to 3,500 m$\mu$ (58).

The wavelength response of a photocell depends on the type of photo cathode used and it is thus important to select the right photo tube for the light source in use. Spectral response curves for various detectors and light sources are shown in figure 3.2.

### Photomultipliers

The current output from the vacuum photocell is very small, and it is necessary to pass it through a very high resistance, (perhaps as high as 2,000,000,000 ohms), in order to develop sufficient voltage to be fed into an amplifier.

Another means of detecting the light after it has passed through the sample is by means of a photomultiplier. This electronic device has a photo cathode and an anode, the electrons leaving the cathode being directed towards an electrode structure called a dynode where secondary electrons are produced. These secondary electrons, are in turn directed, by means of the electric field, to another dynode where the process is repeated. The photomultiplier may have up to 13 dynodes, so that considerable electron multiplication is obtained, before reaching the anode. The spectral response of the photo-multiplier depends on its photo cathode composition and is indicated by an S number designation. Typical characteristics of photomultiplier tubes used in spectrophotometry are listed in table 3.2. All these tubes have an output current which is a linear function of the exciting energy under normal operating conditions.

The usual way of supplying the individual dynode positive potentials, is from a resistance divider chain, and a high voltage stabilised supply. The dynode resistor current should be at least five times the mean current for satisfactory operation, the anode current being limited to the low value of ten microamps for maximum stability of gain. The basic requirement in spectrophotometry is for a photomultiplier with good photo sensitivity and low dark current. The dark current is due to thermal emission from the photo cathode and it is dependent on the cathode area and material (52).

A summary of light sources and detectors with the wavelength range covered is given in table 3.3.

# LIGHT SOURCES AND DETECTORS

| Photo-multiplier | Spectral Response | Wavelength for maximum response (approx.) | Spectral range (approx.) | Photocathode | Window | Equivalent Dark Current input (Lumens) | No. of Stages | Typical Luminous cathode sensitivity μA/Lumen | Used on |
|---|---|---|---|---|---|---|---|---|---|
| 1P21 | S4 | 400 mμ | 300-680 mμ | CsSb | Lime glass | $5 \times 10^{-10}$ L* | 9 | 40 | Eppendorf photometer (fluorimeter attachment). Baird atomic fluorispec |
| 1P28 | S5 | 340 mμ | 190-670 mμ | CsSb | U.V glass | $1.25 \times 10^{-9}$ L* | 9 | 40 | Cary model 14, 15, Beckman DB, DK. Bausch & Lomb precision spectrophotometer. Zeiss PMQII. Hilger spectrochem, gilford, Ultrascan. Optica CF4. Baird atomic fluorispec |
| 931A | S4 | 400 mμ | 300-680 mμ | CsSb | Lime glass | $2.5 \times 10^{-9}$ L* | 9 | 40 | Eppendorf photometer (fluorimeter attachment) |
| 1P22 | S8 | 365 mμ | 290-780 mμ | CsBi | Lime glass | $3.75 \times 10^{-7}$ L* | 9 | 3 | Phoenix Light Scattering photometer (can also use 1P21, 1P28, 931A) |
| 4473 | S4 | 400 mμ | 300-680 mμ | CsSb | Lime glass | $5 \times 10^{-10}$ L* | 9 | 40 | Warner & Swasey 501 Rapid Scanning Spectrometer |
| 6217 | S10 | 450 mμ | 320-800 mμ | Ag-Bi-O-Cs | Lime glass | $2.5 \times 10^{-8}$ L* | 10 | 40 | Cary 14 Scattered Transmission accessory, can also use 7664 (185-600 mμ) & 6292 (300-600 mμ) detectors |
| 9660B | S4 | 380 mμ | 200-680 mμ | CsSb | U.V glass | $1.1 \times 10^{-12}$ L† | 9 | 35 | Unicam SP90 (200-650 mμ) |
| 9663B | S10 | 420 mμ | 180-720 mμ | Bi-Ag-O-Cs | U.V glass | $1.8 \times 10^{-12}$ L† | 9 | 25 | Unicam SP90 (210-770 mμ) |
| 9529B | S10 | 420 mμ (2nd peak at 210 mμ) | 165-750 mμ | Bi-Ag-O-Cs | fused silica | $5.1 \times 10^{-12}$ L† | 11 | 45 | Unicam SP3000, SP900 |
| 6256B | S13 | 400 mμ (2nd peak at 200 mμ) | 165-650 mμ | Cs-Sb-O | fused silica | $9.3 \times 10^{-14}$ L† | 13 | 70 | Unicam SP700 |
| 6255B | S13 | 400 mμ (2nd peak at 200 mμ) | 165-650 mμ | Cs-Sb-O | fused silica | $3.8 \times 10^{-13}$ L† | 13 | 70 | Unicam SP800 |

Notes: It should be noted that the data on photomultipliers is general only and the tubes used by spectrophotometer manufacturers may be specially selected (e.g. for low dark current)

\* R.C.A. data. Refers to maximum equivalent anode dark current input at 25°C.
† E.M.I. data. Refers to typical (wavelength peak) dark current noise equivalent input.

TABLE 3.2.
*PHOTOMULTIPLIERS USED IN SPECTROPHOTOMETRY*

| Instrument | Light Source | Detector | Wavelength Range |
|---|---|---|---|
| *Filter Instruments* | | | |
| Hilger Spekker absorptiometer | Tungsten | Barrier layer photocell | 8 filters between 430 and 680 m$\mu$ |
| Eppendorf Photometer | Mercury Mercury-Cadmium | Photocell | Filter between 313 and 1014 m$\mu$ |
| Perkin Elmer 141 Polarimeter | Sodium. Mercury | Photomultiplier | 5 filters between 313 and 578 m$\mu$ (Hg). 1 filter 589 m$\mu$ (Na) |
| Eppendorf fluorimeter attachment | Mercury Mercury-Cadmium | Photomultiplier | Primary and secondary filters |
| *Monochromator Instruments (Visible and U.V)* | | | |
| Unicam SP600 Series 2 | Tungsten | Red and blue photocells | 335-1000 m$\mu$ |
| Hilger & Watts Spectrochem | Tungsten | Photomultiplier | 340-750 m$\mu$ |
| Unicam SP500 Series 2 | Tungsten. Deuterium | Red and blue photocells | 186-1000 m$\mu$ |
| Beckman DU2 | Tungsten. Deuterium | Photomultiplier Red photocell | 190-1000 m$\mu$ |
| Optica CF4 | Tungsten. Deuterium | Photomultiplier Red sensitive photomultiplier | 185-750 m$\mu$ 185-1000 m$\mu$ |
| Beckman DB | Tungsten. Hydrogen | Photomultiplier | 205-770 m$\mu$ |
| Perkin Elmer 137 | Tungsten. Deuterium | Photomultiplier | 190-750 m$\mu$ |
| Unicam SP3000 | Tungsten. Deuterium | Photomultiplier | 175-750 m$\mu$ (with nitrogen purge) |
| Beckman DK2A | Tungsten. Hydrogen | Photomultiplier lead sulphide cell | 185-3,500 m$\mu$ |
| Zeiss DMR21 | Tungsten. Hydrogen | Photomultiplier lead sulphide cell | 185-2,500 m$\mu$ |
| Cary 14 | Tungsten. Hydrogen | Photomultiplier lead sulphide cell | 186-2,650 m$\mu$ |
| Unicam SP700A | Tungsten. Deuterium | Photomultiplier lead sulphide cell | 187-3570 m$\mu$ |
| Durrum Gibson Stopped Flow | Tungsten Iodide | Photomultiplier | 350-800 m$\mu$ |
| Bellingham & Stanley Polarmatic 62 spectro-polarimeter | Xenon | Photomultiplier | below 200-600 m$\mu$ |
| Cary 60 Polarimeter | Xenon | Photomultiplier | 185-600 m$\mu$ |
| Aminco Bowman Spectrophotofluori-meter | Xenon Xenon-Mercury | Photomultiplier Red sensitive photomultiplier (7102) | 200-800 m$\mu$ 800-1200 m$\mu$ |
| Baird Atomic Fluorispec | Xenon | Photomultiplier | 220-700 m$\mu$ |
| Techtron AA4 atomic absorption spectrophotometer | Hollow cathode lamp. Gas discharge lamp. | Photomultiplier | 186-1000 m$\mu$ |
| Unicam SP90 atomic absorption spectrophotometer | Hollow cathode lamp | Photomultiplier Special photomultiplier | 190-770 m$\mu$ 190-852 m$\mu$ |

| Instrument | Light Source | Detector | Wavelength Range |
|---|---|---|---|
| *Monochromator Instruments (Infra-red)* | | | |
| Unicam SP200 | Nernst filament | Golay cell | 2-15.4 $\mu$ |
| Unicam SP1200 | Nernst filament | Golay cell | 2.5-25 $\mu$ |
| Hilger & Watts Infrascan | Nichrome wire | Vacuum thermopile thermopile | 2.5-15.4 $\mu$ |
| Perkin Elmer 125IR | Globar | Golay cell | 1-25 $\mu$ |
| Beckman IR8 | Nichrome wire | Lead sulphide cell. Evacuated thermocouple | 2.5-16 $\mu$ |
| Beckman IR9 | Nernst filament | Evacuated thermocouple | 2-25 $\mu$ |
| Beckman IR11 | Mercury Arc | Golay cell | 12.5-300 $\mu$ |
| Beckman IR12 | Nernst filament | Evacuated thermocouple | 2-50 $\mu$ |

TABLE 3.3.
LIGHT SOURCES AND DETECTORS USED ON
SPECTROPHOTOMETRIC INSTRUMENTS

# SINGLE BEAM ABSORPTION INSTRUMENTS

### The null balance circuit

A circuit, showing how the photocell and amplifier are connected as a measuring circuit in the Unicam SP500 and Beckman DU spectrophotometers, is given in figure 3.3. The output current of the photocell, proportional to the intensity of the light falling on it, is measured by balancing the voltage drop the current develops across the high ohmic value resistance R1, against the voltage on a calibrated potentiometer T. This potentiometer which is calibrated linearly in per cent transmission is adjusted until a null or zero reading is obtained on the microammeter. The per cent transmission can then be read directly from the scale. Before measuring the sample, it is first necessary to back off the current that flows under dark conditions, by means of a potentiometer DC. The sensitivity of the circuit is varied by adjusting the voltage across the calibrated slide wire potentiometer T, with a multiturn potentiometer S. The bias batteries shown may be replaced by a transformer and rectifier giving a d.c. voltage which is then stabilised by zener diodes. The transistorised circuits used in the Unicam SP500 (series 2) are shown in figure 3.1.

### Photomultiplier circuits

In the photomultiplier circuit of figure 3.4, the feeble photoelectric current produced at the cathode may be amplified to the order of 1,000,000 times by the electron multiplication of the dynode system. The output anode current of the photomultiplier operates the coil of a galvanometer which deflects a mirror attached to its movement, so that a light spot on a scale calibrated in optical density and per cent transmission is also deflected. The gain or range of the system can be varied by switching the high voltage applied to the photomultiplier. This technique is used in the Hilger and Watts Spectrochem instrument.

All the methods outlined so far have given a visual indication which has a linear relationship with the transmission of light through the sample. The potentiometer or galvanometer scale may also be

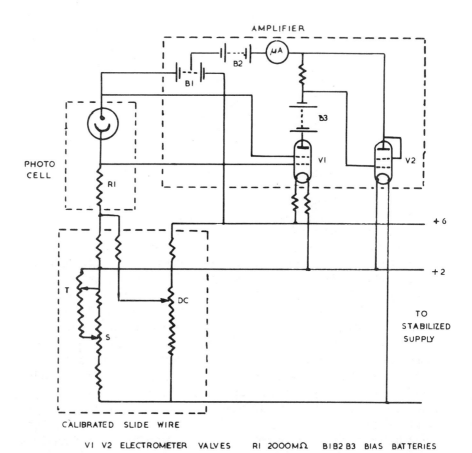

*Fig. 3.3.* Simplified spectrophotometer measuring circuit Unicam SP500 Beckman DU.

calibrated in optical density units, but as the relationship between per cent transmission and optical density is logarithmic the scale becomes cramped above an optical density of 1.0.

### Linear absorbance readout scales

To obtain an output which bears a linear relationship to the optical density, it is necessary to use a special photomultiplier circuit or logarithmic amplification of the photocell output. With a photomulti-

plier detector, a series control or regulating valve may be used to
maintain the anode current at a small but constant value, the change
in the dynode supply voltage necessary to achieve this, being approximately proportional to the optical density of the sample. In figure
3.5, it can be seen that the high voltage supply is divided between
the control valve V1 and the dynode resistance chain. When the
intensity of the light impinging on the photomultiplier tube increases, the anode current tends to increase, so that the voltage
developed across resistance R1 rises. This changes the input voltage
to the amplifier V2, which controls the bias voltage on the control
valve grid, so that more volts are dropped across the control valve
and less across the dynode chain. The effect of reducing the dynode

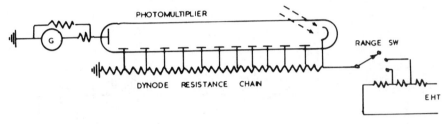

*Fig. 3.4.* Photomultiplier circuit.

voltage is to reduce the gain of the phototube and the anode current
then falls. In this way the anode current is regulated at a constant
level of a few microamps.

The change in the dynode voltage is approximately proportional
to optical density, but if we require a more exact linear relationship, it is necessary to use diodes to shunt the output from the
regulating circuit. The diodes are so arranged that they start to conduct at certain preset voltages, and in doing so they draw just sufficient current to make the voltage fall to the desired value. This
method, first used by Sweet and developed by Gilford for adaption
to spectrophotometers, can give a linear output using eight diodes,
up to an optical density value of 3 or 4 (56) (23) (22).

The Hilger-Gilford linear optical density attachment uses this
technique and may be fitted to many types of single beam absorption
spectrophotometers. The amplifier used in this circuit is of the
cathode coupled type or long tailed pair, where the currents of both
halves of the valve flow through one common cathode resistor. The
amplifier V2 is connected so that if a signal on grid A causes the
current in that half of the valve to increase, the current through
the opposite half decreases. This gives rise to a difference in the anode
voltages and an output signal to V1. Valve V5 is a cathode follower
which connects between the high output resistance of the control
circuit and the low resistance of the output. The output from the

control circuit is balanced against a precision slide wire potentiometer T, (calibrated in optical density units), using a microammeter as a null indicator. It is also possible to connect a recorder across the cathode follower output so that, using the calibrated potentiometer to provide reference voltages corresponding to the background absorbancy of the sample, it is possible to record changes in optical density. This technique is used in the study of kinetic reactions.

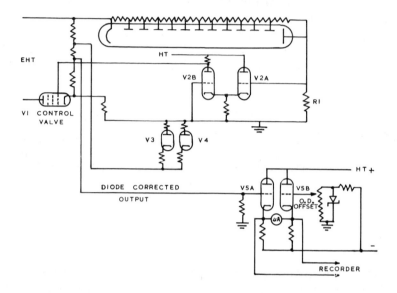

*Fig. 3.5.* Basic circuit for linear optical density output.

## Spectrophotometer modifications

Several circuits have been described in the literature for modifying single beam spectrophotometers for special purposes. For short term recording, it is possible to use the electrometer stage of a Beckman DU amplifier, although this amplifier is not stable enough for long term work (10). Amplifier modifications have been described but an external amplifier may also be used (44) (28). Marr and Marcus feed the photocell output into a Keithley d.c. amplifier and use a silicon diode network at the amplifier output in order to obtain a linear absorbance scale. The diodes are under progressively higher reverse voltages, and this circuit allows linear recording of absorbance at a fixed wave length (38). Kaye and Waska have modified a Beckman DU to scan from 160-210 m$\mu$ in six seconds, this system being used for chromatography column monitoring (35). An automatic recording integrating densitometer modification to the DU has been described by Zak et al (66).

Modifications to the Unicam SP500 have also been described and a recent circuit due to Thompson enables linear optical density recording over the range 0-2, 0-1 and 1-2 OD units. A photomultiplier detector is used, and its anode current produces a volts drop across a switched load resistor feeding an amplifier. The input stage of the amplifier is an electrometer valve which is followed by three transistor stages. Linear to logarithmic conversion is obtained by using a zener diode with a logarithmic change of breakdown voltage for a variation of current through it (57).

**Digital techniques**

Many spectrophotometers are now available with a digital display of absorbance and an example which can have a sample changer unit added and also drive a printer or typewriter, a tape puncher and recorder simultaneously, is the single beam Japanese Shimadzu digital spectrophotometer model AQV-50. In this instrument, full scale, the zero point, and the photomultiplier sensitivity (with a standard cell in the light path) are automatically set on pressing a button. The unknown samples are then measured and the absorbance or transmission value obtained is shown on the digital display and also printed out. The monochromator can be evacuated and measurements made over the range 183-1,200 m$\mu$. Stray light is claimed to be less than 0.1% at 190 m$\mu$ in vacuum.

A block diagram of the instrument is shown as figure 3.6. The photomultiplier sensitivity is automatically adjusted by a system of relays which control the dynode voltage. The relays are operated, so that the difference between the output signal from the photomultiplier and a zener reference voltage is reduced, until the polarity of the voltage difference is reversed. The relay system is then set back one step. With the sample shutter closed the dark current of the photomultiplier is cancelled through the feedback loop via one of the differential inputs of the d.c. comparator amplifier. When measurement is made, the feedback loop is broken and the voltage required for cancellation is maintained by the zero holding amplifier. Similarly, the full scale (100%) holding amplifier, maintains a voltage equivalent to the light signal voltage, and this is fed to the potentiometer when the feedback loop is open and the d.c. comparator is ready for the measurement of unknown samples.

With the sample in position, the light intensity is measured and the difference between the signal and the voltage on the potentiometer slider is amplified by the comparator. This amplifier's output is fed to a transistor gating circuit. When measurement starts, a motor rotates a scaled disc and also the potentiometer slider. The disc, with the lamp and photocell assembly, produces pulses which are fed through the gate to a counting and display unit, providing that the signal from the potentiometer is smaller than the light intensity signal. The gate is closed when the voltage from the potentiometer is larger than the light signal voltage. Counting then stops. The disc has two sets of scales, one linear for transmission and the other

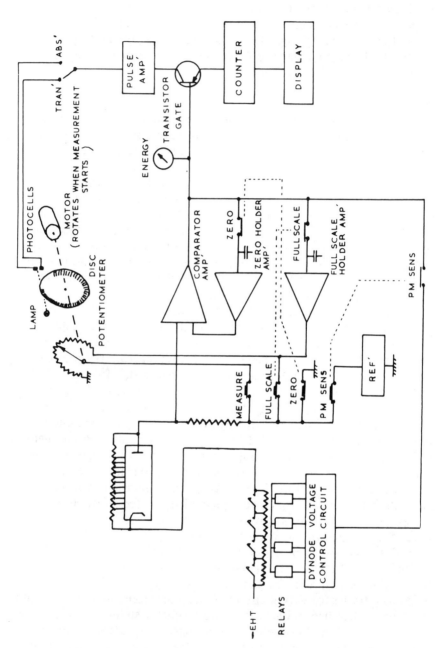

*Fig. 3.6.* The Shimadzu AQV-50 digital spectrophotometer.

logarithmic for absorbance value. For transmission there are 1,000 scalings so that the smallest digit of conversion is 0.1% T.

Separate electronic digital voltmeters, which have a high input impedance, can be used as readout devices on spectrophotometers. A modification to the Hitachi Perkin-Elmer 139 has been described by Burks. In this circuit, a voltage divider network is connected across the meter series resistance in order to feed the external volt meter (8).

## DOUBLE BEAM ABSORPTION INSTRUMENTS

**Beers and Lamberts laws**

Lamberts law states that when a beam of parallel monochromatic light enters an absorbing medium, at right angles to plain parallel surfaces of the medium, each infinitesimally small layer of the medium decreases the intensity of the beam entering the layer by a constant fraction. Beers law states that the intensity of a beam of parallel monochromatic light decreases exponentially as the concentration of the absorbing material increases.

In practice these laws do not always apply well, as the light from the monochromator isolates a small wave band not a single wave length, and the beam may not be completely parallel. In addition to the narrow wave band of selected light from the monochromator, there is also, a small amount of stray light at other wave lengths due to light scattering in the monochromator. Stray light may be reduced by eliminating unwanted reflections in the design of the instrument and by keeping the optical components clean in use. The effect of the stray light is to give incorrect optical density readings.

**Stray light**

There are a number of techniques by which stray light may be reduced. The best method is to use two monochromators in series so that the exit slit of the first is the entrance slit of the second. The use of two monochromators gives increased resolution but is expensive and takes up considerable space. An alternative is to make the light beam traverse the monochromator several times. In a multipass system, described by Walsh, the second order beam is interrupted by a chopper (60) (61). The detector output is passed through an amplifier, tuned to the chopping frequency, so that stray light which has passed through the system only once (first order light) is not measured.

In the double beam in time instrument (split beam), the light beam is switched alternately between a reference cell and the sample cell, usually by means of rotating or vibrating mirrors. These mirrors alternate the light beam between the cells at a low frequency, perhaps between 10 and 70 Hz, and in addition the light beam itself is chopped by a rotating disc. The output from the photoelectric detector is alternating and normal a.c. amplifiers can be used which are not subject to drift (29) (41) (47).

Stray light is reduced by using a rotating mechanical shutter, as the instrument a.c. amplifier ignores d.c. signals, resulting from non interrupted light such as reflection from walls. More stray light can be expected when working in the U.V. part of the spectrum as short wave length light is scattered more than long wave lengths. The actual effect stray light has, depends on what proportion the selected spectrum is of the total source radiation. The greater this proportion, the less will be the effect of stray light. The response of the photodetector is also important as if we are working at a wavelength where this is low, scattered light may occur at wave lengths where response is greater, so that the output signal may depend more on stray light than on the selected wave length.

**Ratio recording instruments**

Spectrophotometers may work on the optical null principle or as ratio recording instruments, these two methods being shown in figure 3.7. The first diagram, figure 3.7(a), shows the mode of operation of the Beckman DK2A ratio recording instrument. The light beam, before passing through the monochromator, is chopped by a rotating disc and after the monochromator switched alternately between the sample and reference cells by rotating mirrors. The output from the photoelectric detector is amplified and then applied to the output switch. This switch, which works in synchronism with the rotating mirrors, allows the signal to pass to the appropriate demodulator (rectifier and filter unit), the switch common contact earthing the input to the other demodulator. This means that if the sample light beam is falling on the photomultiplier detector, then the signal passes through the amplifier to the sample demodulator, the input to the reference demodulator being earthed during this part of the switching cycle.

If a difference exists between the reference and sample, due to light being absorbed in the sample, then a d.c. error signal is produced at the demodulator outputs and a signal is fed into the pen servo amplifier. The output from this amplifier operates a motor, which is coupled to the recorder pen and a potentiometer at the reference channel demodulator output. The motor drives this potentiometer (and also the pen) so that the input to the servo amplifier is reduced to zero. In this way, the output from the sample and reference demodulator channels to the recorder are made equal, and the pen then records the difference in absorbance between the reference and sample. The energy of the monochromatic light beam is maintained at a constant level by another servo system which

DOUBLE BEAM ABSORPTION INSTRUMENTS 99

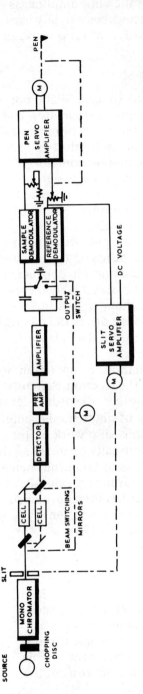

Fig. 3.7. (a) Diagram of Beckman DK2A Ratio Recording instrument.

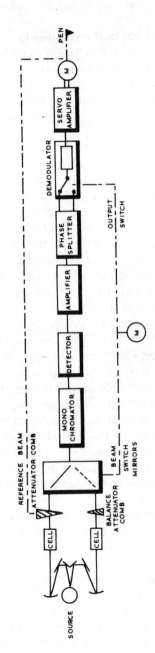

Fig. 3.7. (b) Diagram of UNICAM SP200 Optical Null instrument.

operates the slits in the monochromator. As the same amplifier is used for both sample and reference beam measurements, any variations in the system affects both beams equally, the ratio between the two remaining the same.

## Optical null instruments

Figure 3.7(b) shows a block diagram of the Unicam SP200 double beam spectrophotometer, which utilises an optical attenuator, and works on the optical null principle. The SP200 is a recording infrared instrument, the light from the source being split into two beams, and passed through the beam switch and the monochromator. When the sample and reference beams are unbalanced, the a.c. output from the photocell of the golay detector is amplified and passed through a phase splitter, which produces two outputs 180° apart electrically (that is one output is negative going while the other is positive going). These are passed through the output switch to produce a full wave rectified d.c. signal, which depends on the phase of the a.c. signal from the detector, this phase relationship depending on which beam has the greater energy. The d.c. signal is then converted to mains frequency a.c. by a vibrator chopper, and further amplified to operate the pen servo motor, which drives the recorder pen and the reference beam attenuator comb until the balance between the beams is restored.

These two examples illustrate the basic methods of operation of double beam spectrophotometers, although the electronic circuits used vary from one type of instrument to another. Recording densitometers, used for the measurement of density of photographic emulsions or plates and for scanning chromatograms, also work on the optical null principle. The photomultiplier is usually operated so that its current is controlled at a constant value, the signal output being taken off the dynode resistance chain as we outlined previously. A transistorised circuit for a linear densitometer has been developed by Gordy, Hasenpusch and Sieber which has an optical density range from zero to four with a linearity of ± 0.5 per cent (26).

## Photomultiplier operation

Spectrophotometer detector and cell compartments may be constructed so that the photomultiplier is subjected to high incident light radiation when the cell compartment cover is raised, and if this occurs when voltage is applied to the dynode, the tube can easily be saturated and possibly damaged. To overcome this it is usual to (1) provide a shutter which closes whenever the sample compartment cover is raised or to (2) automatically control the sensitivity of the photomultiplier by turning off the EHT voltage. Also, in order to minimise the effects of stray magnetic fields, the photomultiplier may be enclosed in a mu-metal screen. Illumination from the reference and sample beams should fall on identical parts of the photo sensitive surface, as the spectral sensitivity of the surface is rarely uniform. In order to avoid output balance being critically dependent on the exact position of the detector, a diffusing screen or a lens focusing on one part of the surface may be used.

## Signal amplification

A pre-amplifier is used in close proximity to the detector to amplify the small signals to a level when they can be passed to the main amplifier unit. Within the pre-amplifier, extra shielding and earthing precautions are necessary, or transient noise voltages from inductive or capacitive coupling may obscure the signal. The photomultiplier is often followed by a cathode follower as the photomultiplier output is at a very high resistance. The cathode follower circuit has a high input resistance and low output resistance, so that loading of the photomultiplier output is reduced, and a low impedance path is provided for the electronic circuits which follow.

The main amplifier is of the conventional resistance capacitance coupled type with negative feedback.

## System performance

It is essential that the beam switch rotating sector mirrors are synchronised with the switching carried out by the output switch prior to demodulation, and this is achieved by either operating the switch contacts by a mechanical drive common to the optical beam switch, or by the use of electronic circuits.

The use of a chopping disc, to modulate the light beam at a frequency higher than the beam switching frequency, provides an optimum signal to noise ratio. Dynode voltage control of the photomultiplier changes its sensitivity. Increasing the sensitivity increases noise, and the noise appearing on the recorded spectrum is dependent on both the settings of the sensitivity control and the monochromator slits. The signal output of a photomultiplier increases linearly with the light level but noise increases as the square root of the light level. In a system where all the other noise sources are smaller than the photo tube noise (shot noise limited), the noise level increases with the resolution (spectral band width). Since the energy leaving the monochromator is proportional to the square of the slit width, and the phototube noise is proportional to the square root of the incident light level, system noise is proportional to slit width. Where thermal noise (e.g. from a lead sulphide cell) is the limiting factor, noise increases with the square of the resolution. It should be noted, that the noise level from a photo conductive cell, is not affected by changes in the incident light level (9).

The input to the demodulator, which rectifies or converts the a.c. signal to a d.c. voltage, may be tuned to the chopping frequency by means of resistance and capacitance parallel T networks or in-

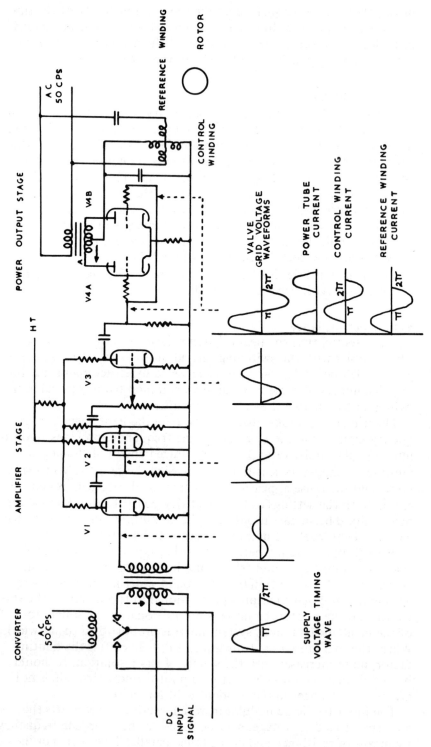

Fig. 3.8. Basic Servo amplifier circuit.

ductors which have a high Q value. These circuits reject all frequencies not close to the chopping or carrier frequency and help to eliminate spurious or random noise signals.

**The recorder servo amplifier**

In the recorder servo amplifier, the d.c. signal is converted to a.c. (for stable amplification), by a chopper or converter working in conjunction with a special transformer. The converter may consist of a flat metal reed which oscillates between two contacts (figure 3.8), the oscillation usually being at mains frequency by the action of an energising coil. The converter causes the d.c. input to be passed through the two halves of the transformer winding alternately, this giving rise to an alternating flux in the transformer core and so to an alternating voltage in the secondary winding. If the polarity of the d.c. signal is reversed then the current flow in the transformer reverses and the output wave form is 180° out of phase with the previous case.

The a.c. amplifier used has normal resistance capacitance coupling, the power output stage operating in the following way. If we consider the first half cycle $0$-$\pi$ of the mains voltage then from figure 3.8 we see that with a d.c. input signal to the converter the a.c. signal on V4a grid is in phase with the supply. At this time, end A of the secondary winding of the output transformer, which is also energised from the mains, will be positive so valve V4a conducts and current flows through the motor winding as shown by the solid arrow. In the second half $\pi - 2\pi$ of the mains cycle, the power transformer T2 secondary polarity is reversed, so V4b anode is positive, but the valve does not conduct since the signal on its grid is now negative. If the d.c. input polarity to the converter is reversed we have an a.c. signal at V4 grid which is 180° out of phase with the supply voltage timing wave, the effect of this being to cause V4b to conduct since its grid is positive during the period $\pi - 2\pi$.

The motor control winding and capacitance C form a parallel circuit, the current which flows in the motor winding being 90° out of phase with the power tube current. The motor used is of the two phase induction type, the direction of rotation depending on the relative phase relationship of the currents in the control and reference windings. The reference winding and capacitance form a series circuit, the current being in phase with the supply. Thus for an "in phase" input to the power amplifier, the current in the control winding lags the reference winding current by 90°, as shown in the wave forms of figure 3.8. If the d.c. input voltage polarity changes then the control winding current leads the reference winding current by 90°.

Rotation of the motor actuates the recorder pen and the wiper of the balancing potentiometer. The chart drive of the spectrophoto-

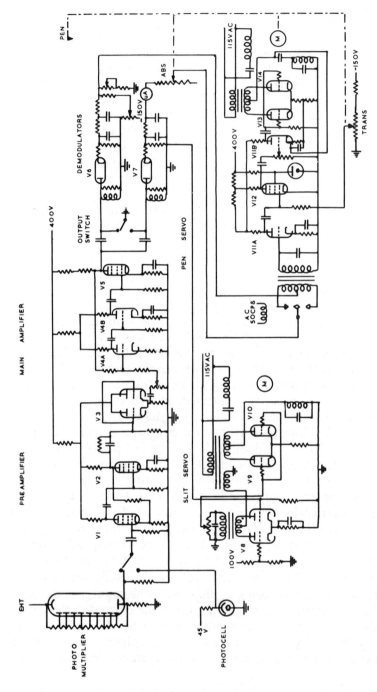

*Fig. 3.9.* Simplified circuit of Beckman DK2A. It should be noted that switches and valve screen voltages have been omitted and the filter section of the demodulators and servo amplifier power output stages simplified.

meter recorder is usually a synchronous motor, which may work in conjunction with gears to give a constant speed time drive at a fixed wavelength, or the chart may be coupled to the prism wavelength drive so a plot of wavelength against absorbance or transmission is obtained.

In the servo system it is necessary to incorporate some form of damping as oscillation or hunting may occur. This means that the mechanical output and balancing potentiometer, in moving to reduce the amplifier input to zero, overshoots the zero point due to inertia in the system and so gives rise to a signal causing the motor to reverse. This sequence is repeated so that the system oscillates about the null position. To overcome this difficulty, we can adjust the sensitivity or gain of the system to the required level by means of a potentiometer. This is usually done by increasing the sensitivity control so that the system just starts to oscillate, and then reducing the setting of the control until the oscillation just stops. Another method of damping is velocity feedback using a small generator driven by the balancing motor. In this system, a voltage proportional to the velocity of the balancing motor is fed back to the amplifier input.

Having considered the basic electronic components we will outline how they are used in various instruments, the spectrophotometers chosen being those of which the author has personal knowledge and which show the fundamental operational methods.

### The Beckman DK2A ratio recording instrument with servo control of monochromator slits

A simplified circuit diagram is shown as figure 3.9, the detector output passing via the preamplifier V1 V2 and cathode follower V3, to the main amplifier comprising V4a, V4b, V5. From here the signal is passed through the rectifiers (V6 or V7) and filters comprising the demodulator section, the output of which feeds the pen servo amplifier.

The rectified output voltage from the reference channel is also compared against a fixed reference voltage at V8, any difference causing the slit servo motor to operate and adjust the monochromator slit to keep the energy output from the light beam constant. Metering is also provided for the reference energy level.

When making absorbance measurements, it is necessary to vary the pen servo amplifier gain, to ensure equal sensitivity over the absorbance scale. This is because the absorbance slide wire is logarithmic, so that 1/10th in. wiper travel at one end produces a larger voltage error than 1/10th in. travel at the other end. The amplifier gain is varied by altering the bias on valve V12, the bias voltage being obtained from the transmittance slide wire wiper which is mechanically slaved to the absorbance slide wire.

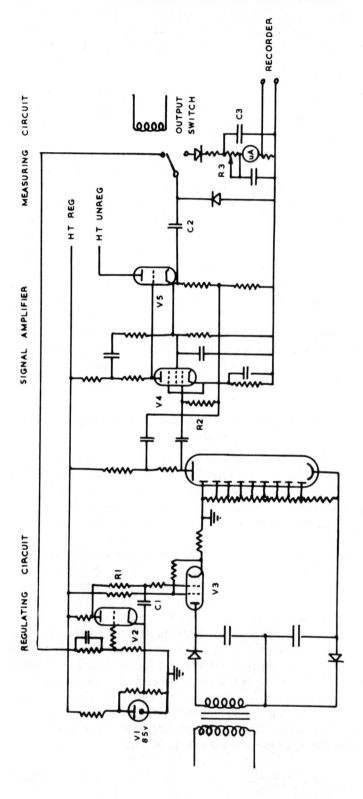

*Fig. 3.10.* Simplified circuit of Beckman DB.

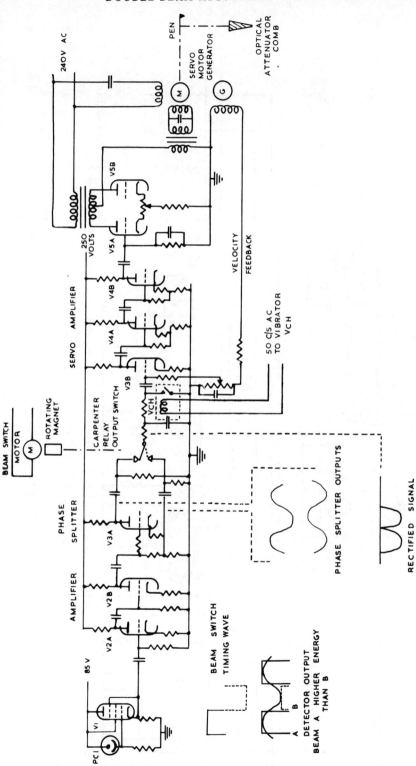

Fig. 3.11. Basic circuit of UNICAM SP200 (waveforms idealized for circuit explanation).

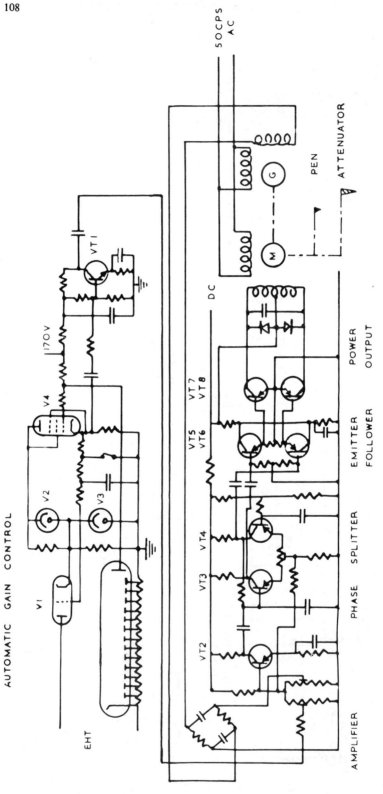

*Fig. 3.12.* Simplified circuit of UNICAM SP 800.

## The Beckman DB ratio recording instrument with automatic photo-multiplier sensitivity control of reference level

This instrument utilises a different method of maintaining the reference level at the required constant value, there being no servo control of the slits.

The reference signal is compared with a stable reference voltage in a regulating circuit, any difference causing the photomultiplier sensitivity to change until the reference signal again equals the correct value. The circuit is shown as figure 3.10. Part of the reference beam signal is compared with a reference voltage on the cathode of V2, and if they are equal, V2 conducts and capacitor C1 is discharged through resistance R1. This causes the voltage supply to the grid of control valve V3 to be reduced, and the valves effective resistance in the dynode control circuit is such that the current flowing produces a voltage on the dynodes that is just sufficient to maintain the reference signal at the desired level. If for any reason the level of the reference signal changes, then the bias voltage on valve V3 changes the current through the dynode resistance chain, thereby adjusting the sensitivity of the photomultiplier so that the reference is maintained at the correct value.

The light beam is alternated between the reference and sample at 35 Hz by vibrating mirrors, and collected electrons at the photo-multiplier anode cause a negative going voltage on the grid of V4, decreasing the current through that tube, and causing the anode voltage to go in a positive direction. In V5, the cathode volts follow the positive going voltage applied to its grid, and are coupled by capacitance C2 and the output signal switch, to either the regulating circuit or the metering circuit depending on the switch position. Part of the output of valve V5 is fed back to the grid of V4, through resistance R2, to stabilise the gain of the amplifier to approximately 20. The beam switching assembly consists of a pair of parallel mirrors, which are caused to vibrate with simple harmonic motion by an electronic unit. This unit also provides a signal to drive the magnetic mercury contact output switch, which is thus synchronised with the vibrating mirror system.

## The Unicam SP200 optical null instrument with valve amplifier and Golay detector

In this spectrophotometer (figure 3.11) the a.c. 10 Hz output from the photocell of the Golay infra red detector is connected through a cathode follower stage V1 to the conventional amplifier, comprising valves V2a and V2b, and then to a phase splitter V3a. The outputs of V3 pass through the output signal switch, which is magnetically coupled to a beam switching motor, so that a full wave rectified d.c. signal is produced at the common contact. This d.c.

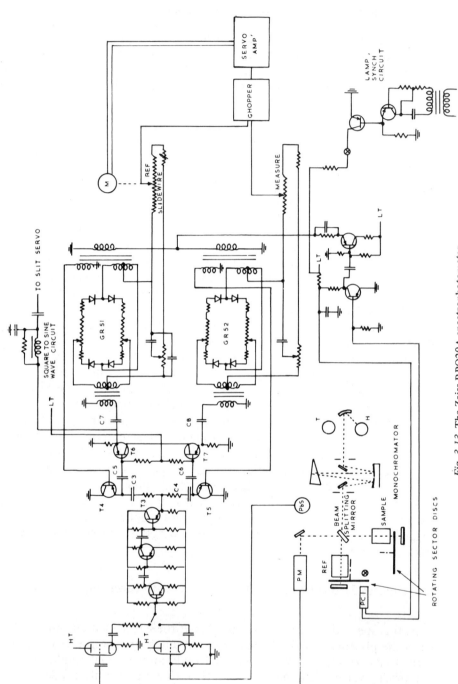

*Fig. 3.13.* The Zeiss RPQ20A spectrophotometer.

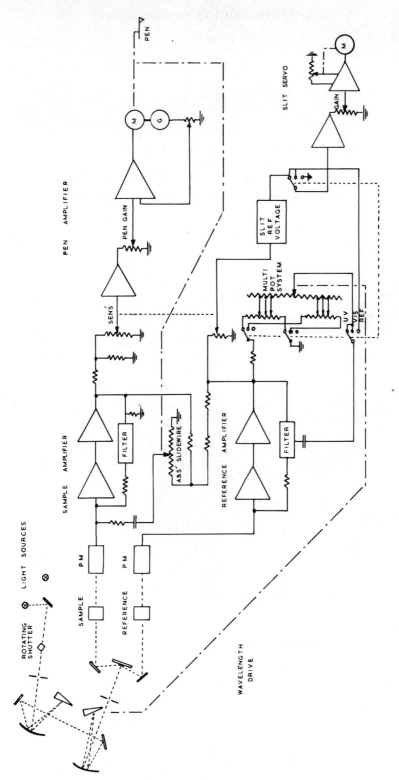

Fig. 3.14. The Cary model 15 double beam spectrophotometer.

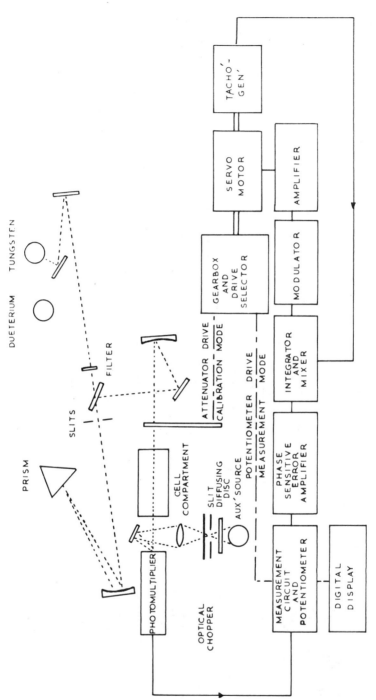

Fig. 3.15. The automatic digital Unicam SP 3000 spectrophotometer.

signal is now converted to 50 Hz by a vibrator chopper and is amplified by valves V3b V4a and V4b. Valves V5a and V5b operate a servo motor which is coupled to the optical attenuator and recorder pen. The attenuator is inserted into the light beam until both beams are of equal energy, the servo system being damped by velocity feedback.

### The Unicam SP800 optical null instrument with automatic gain control of the photomultiplier

In this instrument (figure 3.12) the light beam is switched by a two vane mirror. The detector is a 13 stage E.M.I. photomultiplier and any difference in intensity between the sample and reference beams produces a 50 Hz error signal. The voltage applied to the photomultiplier dynode resistance chain is varied so as to keep the output constant, this automatic gain control ensuring correct pen response throughout the absorbance range of 0 to 2.

The 50 Hz output from the cathode follower V4 is passed through two transistor amplifier stages to two transistors VT3 VT4, connected as a long tailed pair and working as a phase splitter. The two outputs, 180° out of phase, are passed through emitter followers VT5 and VT6 and on to the power output transistors VT7 and VT8. These transistors feed the centre tapped control winding of the pen servo motor which drives the attenuator combs in such a direction as to reduce the difference in intensity between the two beams. Velocity feedback to the amplifier input from a generator coupled to the pen motor provides damping for the attenuator and pen.

### The Zeiss RPQ20A Recording Spectrophotometer with electronic switching circuits for beam splitting and demodulator synchronisation

Electronic switches may be used instead of switched electrical contacts for demodulation. This technique is used on the Zeiss RPQ20A recording spectrophotometer and a simplified circuit is shown as figure 3.13. The output from the photomultiplier or lead sulphide detector is connected to a transistor amplifier through a cathode follower circuit. The amplifier feeds an emitter follower stage and the signal developed across the emitter resistance is a sequence of both measuring and reference pulses. A motor drives sector discs in the monochromator, to separate the reference and sample beams, and also causes control pulses to be generated in the lamp and photocell circuit PC1. These pulses control the switching transistors T4 and T5 which are switched by square waves 180° out of phase with each other. With electronic switch T5 closed, and during the dark interval on this measuring channel, capacitor C4 assumes a potential Vd. When T5 is open, upon arrival of the measuring pulse (Vd + Vm), only the measuring signal Vm appears across it since C4 still retains a voltage equal to and of opposite polarity to Vd. Consequently T5 passes only measuring pulses to C6 and output transistor T7. Operation is similar for the reference channel, T4 being open when T5 is closed. The pulses are fed via transformers to ring rectifier circuits GR51 and GR52. The reference voltage for the rectifier circuit is obtained from the pulse generator. The two rectifier branches conduct alternately and a d.c. output voltage is obtained which is proportional to the amplitude of the appropriate light pulse. These d.c. reference and measure voltages are then fed to a servo amplifier recorder circuit. The instrument also has servo control of the monochromator slits.

## The digital unicam SP3000 spectrophotometer

The recently introduced automatic Unicam SP3000 (figure 3.15) does not use the technique of splitting one beam into two to pass through the reference and sample cells but instead uses one beam and alternatively moves the cells into it. While not a double beam spectrophotometer it is included here as it is a precision instrument that uses an auxiliary standard beam. This beam is derived from a separate tungsten source and is modulated to a 50 Hz square wave form by a synchronous rotating shutter. In the automatic programme, the reference cell is first moved into the monochromatic light beam and the instrument is automatically calibrated for 100% transmission by an attenuator disc, which adjusts the radiation transmitted by the reference cell until it is equal in intensity to the auxiliary standard beam. The sample cell is next moved into the main beam, and the light transmitted through it is measured as a ratio of the auxiliary beam intensity. An E.M.I.9529B photomultiplier is used to measure the light intensity and the measuring circuit is of the null balance type, employing a high accuracy servo operated potentiometer. A coded disc is attached to the potentiometer spindle and converts the potentiometer angular position into a digital result as per cent transmission or absorbance.

The sample changer of the instrument holds 50 samples and measurements may be made at 5 preselected wave lengths (10 with an accessary) within the range 175-750 m$\mu$. Response time is 30 seconds for automatic repeat measurements. A digital printer accessary enables the transmission or absorbance value and an index number to be printed out.

The continuous standardisation with the auxiliary radiation standard ensures that the ratio of the two measured beams remains constant and errors due to instrumental variations such as ageing of lamps are eliminated.

## Special circuits

Few modifications to commercial double beam spectrophotometers are described in the literature but one of interest to biochemists is that described by Dixon (18). The circuit described is an extinction (absorbance) scale expansion unit for the Beckman DK2A spectrophotometer and enables any 0.1 OD portion of the scale to be expanded and applied to give full scale deflection on a slave recorder.

The digital techniques described for single beam instruments can of course be used on split or double beam spectrophotometers. Peterson et al. have described the conversion of the Perkin Elmer 521 infra red double beam optical null instrument for digital recording of spectra. A photoelectric reader system operates a trigger circuit at each calibration mark of the wavelength dial when a digital voltmeter reads the voltage across the transmission slide wire (46).

## DOUBLE MONOCHROMATOR INSTRUMENTS

### Split beam and dual wavelength operation

In conclusion we should mention instruments which have two monochromators or a twin diffraction grating arrangement. This type may be used as (1) a conventional split beam instrument with both monochromators set at the same wavelength, the relative transmission or absorbance between the beams being recorded while the wavelength is varied or (2) as a dual wavelength instrument. In the dual wavelength mode one monochromator is set at a wavelength for unchanging transmission in the sample under examination and this beam used as a reference. The other monochromator is set for a wavelength that is selective to a compound that will undergo changes in the sample. The two beams are initially balanced so that any change from the initial state of equilibrium causes an unbalance, the two beams becoming of unequal intensity.

The Phoenix Precision Instrument Company dual wavelength spectrophotometer combines both these modes of operation and when used as a dual wavelength instrument the alternating measure and reference beams are caused to converge on a single cuvette. The difference in intensity between the two beams is detected by a photomultiplier, the signal being amplified by two amplifiers when in the measure part of the operating cycle. The first amplifier is a difference amplifier while the second has a logarithmic transform for optical density conversion, its output feeding an XY recorder. In the reference portion of the cycle the photomultiplier signal is compared against a stable reference voltage, any difference causing the multiplier dynode voltage to change so that the reference signal is maintained at the correct level. This dynode feedback control standardises the instrument some 60 times a second and ensures that the sensitivity of the spectrophotometer does not change with time.

# SPECTROPHOTOMETRIC INSTRUMENTS

## Single Beam Null Balance Instruments (Manual)

| Range | Optical system | Light sources | Detectors | Photometric accuracy | Photometric reproducibility | Stray light less than | Resolution | Example |
|---|---|---|---|---|---|---|---|---|
| 335-1000 mμ | Littrow silica prism | Tungsten or quartz iodine | Red and blue photocells | ± 0.5% T | ± 0.5% T | 1% | 3 mμ | Unicam SP 600 Series 2 |
| 186-1000 mμ | Littrow silica prism | Tungsten (320-1000 mμ) Deuterium (186-320 mμ) | Red and blue photocells | ± 0.3% T (direct meter readout facility, less accurate) | ± 0.1% T | 0.2% at 200 mμ | 1 mμ (400-800 mμ) 0.5 mμ (200-400 mμ) | Unicam SP 500 Series 2 |
| 185-1000 mμ | Littrow grating | Tungsten Deuterium | Photomultiplier (185-750 mμ) Red photocell (600-1000 mμ) | ± 0.2% T | ± 0.1% T | 0.2% at 210 mμ | 0.1 mμ (600 mμ) | Optica CF4 |
| 190-1000 mμ | Silica prism | Tungsten Deuterium | Photomultiplier (190-600 mμ) Red photocell (600-1000 mμ) | ± 0.3% T | ± 0.1% T | 0.2% at 210 mμ | 1 mμ (visible) 0.3 mμ (U.V.) | Beckman DU2 (a.c. operated) |

## Single Beam Direct Reading on Linear Absorbance Scale

| Range | Optical system | Light sources | Detectors | Photometric accuracy | Absorbance ranges | Stray light less than | Resolution | Example |
|---|---|---|---|---|---|---|---|---|
| 200-800 mμ | Littrow prism | Tungsten (400-800 mμ) Deuterium (200-400 mμ) | Photomultiplier | ± 0.5% | 0-1 0-3 | 0.3% at 210 mμ | 0.5 mμ (200 mμ) 2.5 mμ (600 mμ) | Optica Denistronic |

## Double Beam in Time: Optical Null Instruments (Recording)

| Range | Optical system | Light sources | Detectors | Photometric accuracy | Photometric reproducibility | Stray light less than | Resolution | Example |
|---|---|---|---|---|---|---|---|---|
| 190-850 mμ | Silica prism | Tungsten Deuterium | Special red sensitive photomultiplier | ± 0.02A | ± 0.005A | 1% at 200 mμ | 0.1-0.4 mμ (190-370 mμ) 0.4-2.0 mμ (370-700 mμ) | Unicam SP 800B 0-2 absorbance range |
| 190-850 mμ | Silica prism | Tungsten Deuterium | Photomultiplier | ± 0.01A | ± 0.005A | 1% at 190 mμ | 0.2 mμ (210 mμ) 1.5 mμ (600 mμ) | Perkin Elmer 402 0-1.5 absorbance range |
| 2-15.4 μ (5000 cm$^{-1}$ to 650 cm$^{-1}$) | NaCl prism | Nernst filament | Golay cell | ± 1% T | ± 0.5% T | 2% over most of range | 6 cm$^{-1}$ (1400 cm) | Unicam SP 200 |
| 2.5-50 μ (4000 cm$^{-1}$ to 200 cm$^{-1}$) | 4 gratings | Nernst filament | Evacuated thermocouple | ± 1% T | — | 1% at 500-200 cm$^{-1}$ 0.5% at 4000-500 cm$^{-1}$ | 0.25 cm$^{-1}$ (923 cm$^{-1}$) | Beckman IR 12 |
| 12.5-300 μ (800 cm$^{-1}$ to 33 cm$^{-1}$) | 4 gratings | High pressure mercury arc | Golay cell | 1% + noise level | — | 0.4% at 200 cm$^{-1}$ 4.0% at 80 cm$^{-1}$ | 0.5 to 1.0 cm$^{-1}$ | Beckman IR 11 |
| 11-35 μ | CsBr prism | Nichrome wire | Evacuated thermocouple | 1% T | — | 1% at 11-20 μ 5% at 35 μ | 6 cm$^{-1}$ (400 cm$^{-1}$) | Beckman IR5A |

# DOUBLE MONOCHROMATOR INSTRUMENTS

TABLE 3.4.
PERFORMANCE OF ABSORPTION SPECTROPHOTOMETER MEASURING SYSTEMS

| Range | Optical system | Light sources | Detectors | Photometric accuracy | Photometric reproducibility | Stray light less than | Resolution | Example |
|---|---|---|---|---|---|---|---|---|
| *Double Beam in Time: Ratio Recording Instruments* | | | | | | | | |
| 185-3,500 mµ | Silica prism | Tungsten Deuterium | Photomultiplier lead sulphide cell | 0.5% T (in U.V.) | 0.2% T | 0.1% at 210-2720 mµ | 0.2 mµ (220 mµ) | Beckman DK |
| 187-3,570 mµ | Prism (187-2500 mµ) grating (2500-3570 mµ) | Tungsten Deuterium | Photomultiplier lead sulphide cell | 0.5% T | 0.25% T | 0.1% over most of range 1% at 200 mµ | 0.12 mµ (250 mµ) 0.0026 mµ (3 µ) | Unicam SP 700A |
| 190-800 mµ | Grating | Tungsten Deuterium | Photomultiplier | 0.005A at 0.5A 0.5% T | 0.2% T | 0.1% at 220 mµ | — | Perkin Elmer 124 direct reading |
| *Double Monochromator: Double Beam Instruments* | | | | | | | | |
| Range | Optical system | Light sources | Detectors | Photometric accuracy | Photometric reproducibility | Stray light less than | Resolution | Example |
| 185-800 mµ | Prism monochromators | Tungsten Deuterium | Sample and reference photo-multipliers (electrical null at amplifier input) | 0.005A at 1A 0.012A at 3.4A | 0.003 at 2.4A | 0.001% at 200-600 mµ 0.1% at range limits | — | Cary Model 15 |
| 1-16 µ | Double 2 NaCl prisms | Nernst filament | Evacuated thermocouple | 1% | — | 0.1% at 14.3 µ | — | Beckman IR4 |
| 2.5-15.4 µ | Grating and NaCl prism | Nichrome wire | Vacuum thermopile | 1% | 0.5% | 1% | — | Hilger & Watts Infrascan |
| *Digital Spectrophotometers* | | | | | | | | |
| Range | Optical system | Light sources | Detectors | Photometric accuracy | Photometric reproducibility | Stray light less than | Resolution | Example |
| 183-1200 mµ | Silica prism | Tungsten Hydrogen | Photomultiplier photocell | ± 0.2% T | ± 0.1% T | 0.1% at 190 mµ (in vacuum) | 0.05 mµ (205 mµ) 0.12 mµ (400 mµ) | Shimadzu AQV-50 |
| 175-750 mµ | Silica prism | Tungsten Deuterium | Photomultiplier | Readout resolution 0.1% T 0.001A (0-1.2A) 0.01A (1.2-2.0A) | ± 1 unit (digit) in readout | 0.1% at 196 mµ | 0.5 mµ to 2.0 mµ over working range | Unicam SP 3000 |

### The Cary model 15

In the Cary model 15 double monochromator instrument shown in figure 3.14, light from the tungsten or hydrogen lamp is chopped at 50 or 60 Hz and passes into the first monochromator. The exit slit allows the desired radiation to enter the second monochromator section, the beam traversing this unit in reverse order to the first. Since the optical elements are similar, the principal aberrations of the mirrors are cancelled out. The dispersed radiation from the second monochromator exit slit passes through lenses on to a fixed beam splitter and then to the sample and reference cells. The beams after passing through the cells are focused by a lens system on to two separate phototubes.

RCA 1P28 photomultipliers are used and are protected by shutters when the cell compartment is opened. The instrument works on the null balance principle. A fraction of the amplified reference photomultiplier signal (determined by the slide wire contact position) balances the signal directly at the sample photomultiplier anode. When an unbalance occurs, the resulting signal is amplified by the sample amplifier and the following pen amplifier so that the pen motor drives the slide wire contact to the balance point. Since the null balance is achieved before amplification, variations in the linearity or gain of the amplifiers do not affect the balance point and will have a minimum effect on accuracy.

Variations in response (spectral sensitivity) between the photomultiplier and cells are compensated by an electrical multipot system coupled to the wavelength drive. Correction is performed at 44 points over the entire range with linear interpolation between points. (The multipot system is also used on other commercial spectrophotometers).

The pen motor is coupled to both slide wire brushes. The density slide wire is a uniform ladder network with close spaced taps so the attenuation varies exponentially with linear slide wire movements. The voltage from the slider is fed to the sample photomultiplier load resistance and phased to oppose voltage developed by the photomultiplier current. The gain of the sample amplifier is automatically increased with increasing absorbance in order to avoid sluggishness at high levels.

Without any automatic slit servo system the 100% reference signal would vary with wavelength. A portion of the reference amplifier output, via the slit sensitivity potentiometer, is compared with an a.c. reference voltage and any difference is amplified, filtered in a synchronous filter, passed through a power output stage and applied to the signal phase of the slit servo motor.

The system has a photometric accuracy of 0.002 to 0.005 in the 0-1 absorbance range and 0.012 at the limit of the 1–3.4 absorbance range. With care in sample handling techniques, an analytical accuracy of 0.2% can be obtained on concentration determination. The standard wavelength range is from 1850 Å to 8,000 Å. The performance of various absorption spectrophotometer measuring systems is summarised in table 3.4.

## FLAME PHOTOMETRY

### Principle

In flame photometry the sample is introduced into a flame as a fine atomised mist or spray, the atoms of the sample absorbing energy so becoming excited. In falling back to the ground state light is emitted, the wavelength of which is characteristic of the element. Sodium and potassium are readily determined using a simple flame photometer. A typical instrument may consist of the burner/atomiser unit, a filter to isolate the emission (characteristic of the element) and a barrier layer cell as the detector. A light spot galvanometer with a period of 3-4 secs is a usual measuring device. A battery and a potential voltage divider circuit may be used to provide a back off voltage to eliminate the background reading due to the flame. The barrier layer cell must be kept cool due to its temperature coefficient. The fatigue effect (less current than normal) after exposure to strong light may also be troublesome. To overcome this the cell must be left in darkness to recover.

### Photocell instruments

Circuits have also been described using two photocells, which can be used for direct measurement, or for measurements using the compensation method utilising an internal standard (27). Photoemissive cells are also used, and the caesium-silver oxide photo cathode is useful, due to its response peaks around 800 m$\mu$ and 370 m$\mu$. A circuit has been described by Warren for a flame photometer for routine biochemical use (64).

### Photomultiplier detectors

In order to determine alkaline earth metals, high flame temperatures are necessary and filter characteristics more stringent, so that a photomultiplier detector and amplifier are used. In flame photometry the sensitivity of photomultiplier is limited by the small area of cathode exposed to light. This has led to the use of lens systems so that the flame image covers a large area on the photo cathode. Image converters have also been used for the measurement of potassium using photomultiplier detectors (42). The degree of non linearity and the response time of a photomultiplier to a new level of illumination have been studied by Alkemade (1). Photomultipliers may be used for direct measurements or two used in a compensation circuit using an internal standard. This method has been described by Heidel et al (31). Examples of photomultiplier amplifier circuits for flame photometry have been given by Ramsay (49) and Brealey and Ross (7).

Heavy metals can be determined using for example an acetylene-oxygen gas mixture giving flame temperatures of 3,100°C. A quartz monochromator is used and attachments for most spectrophotometers are available in which the light source is replaced by an atomiser-burner unit.

**Separate light path instruments**

In instruments designed specifically for flame photometry, there may be several slits around the flame so that separate light paths with filters are available for photocells responsive to say lithium, sodium, and potassium. The Instrumentation Laboratory Incorporated Model 143 flame photometer uses this technique and a two step calibration procedure is adopted. A lithium standard is first atomised and the readout of the sodium and potassium channels are set to read zero. The background current from the detector and the flame are thus cancelled out. A standard solution containing the same lithium concentration as first used, plus a known concentration of sodium and potassium, is then atomised and the readout adjusted so that the expected concentration is indicated. The instrument is now set up to measure samples. The outputs from the photocells are amplified and passed to a mathematical circuit which includes two servo amplifiers. The zero adjustment electronically subtracts the background current from the analogue circuit, the voltage output of which feeds digital displays for sodium and potassium concentration.

Separate photomultipliers may be used for the simultaneous measurement of several radiations and instruments have been described by Vallee (59), Heidel and Fassel (31) and Mitchel (42). Some photomultipliers used for the detection of typical elements are as follows; 1P28 for sodium, magnesium and strontium, 1P21 for calcium and manganese and the 1P22 for potassium.

**Integrating circuits**

In flame photometry, variations in line emission and background may be troublesome when working at high sensitivities. These effects may be overcome by using an integrating circuit which integrates the signal received over a comparatively long period. This method is used on the Southern Analytical A1740 grating flame photometer. Light from the flame passes through the grating monochromator to an aluminised quartz plate, at 45° to the direction of light and on which is ruled the slit. Part of the incident light beam (background radiation on either side of the selected line) is reflected to a "background" photomultiplier, while the selected line passes through the slit to the line photomultiplier. The outputs from the two photomultipliers are amplified, integrated and applied differentially to a meter. The sensitivity of both measuring circuits is adjusted so that the difference display on the meter corresponds to the light emitted from the sample element. The integration time may be set to 2, 5, 10, 15, or 20 seconds and is started by means of a push button. On completion of the set time an indicator lamp comes on and the meter reading is held steady. The circuit can then be reset and another measurement taken. A d.c. background compensation circuit is provided as an alternate to the automatic system. Typical 2 $\delta$ detection limits (in ppm) for this method under normal laboratory conditions quoted by the manufacturer, using automatic background compensation and a 10 second integration time, are calcium (4226 Å) 0.02, potassium (7665 Å) 0.01, lithium (6707 Å) 0.00004, sodium (5890 Å) 0.0001.

## ATOMIC ABSORPTION

### Principle
With a normal air-acetylene flame only about 1% of the atoms in the flame are excited to emission, so that the majority are in a minimum energy condition. With the sample vapourised in a flame the atoms are also chemically unbound and absorb radiation at narrow wavelength bands. In atomic absorption spectrophotometry a hollow cathode lamp is used as the source of radiation (62) (34). The cathode of the lamp is made of the element to be determined and the lamp emits the line spectrum of that element. The sample absorbs only at one line, called the resonance line, so that after passing through the flame the radiation resonance line is diminised. The photodetector sees only this reduced resonance line, as the radiation is passed through a high resolution dispersion monochromator, which removes other lines in the spectrum. The atomic absorption method obeys Beers law closely enough (for most elements) to enable direct concentration readout scales.

### Hollow cathode lamp operation
The hollow cathode lamp is operated so that modulated light is produced, the amplifier following the photodetector being tuned to the same frequency. In this way, the flame background and interference, due to sodium and potassium in the flame, are eliminated since they result in a steady or d.c. signal from the detector. When changing from one element to another, time must be allowed for the hollow cathode lamp to warm up (about 15 minutes). This time

delay may, however, be overcome by using multi-element hollow cathode lamps (53) (59). Commercial examples are the calcium-magnesium and copper-nickel-cobalt-chromium-iron-manganese lamps used with the Perkin Elmer 290 atomic absorption instrument. This spectrophotometer uses a littrow grating monochromator and the tuned amplifier system outlined above. Another method of eliminating the warm up delay is to use a lamp turret holding several single element hollow cathode lamps, which are independently run at the appropriate currents. This technique is used in the Unicam SP90 atomic absorption spectrophotometer.

**The single beam unicam SP90**

In the Unicam SP90 (figure 3.16), the light from the hollow cathode lamp passes through the flame and is focused by mirrors

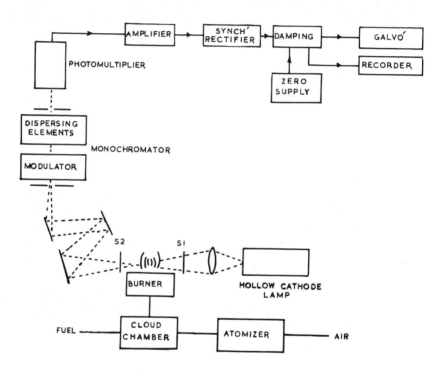

*Fig. 3.16.* Unicam SP90 spectrophotometer.

on to the entrance slit of the silica prism monochromator. The light beam is chopped by the radiation modulator, inside the entrance to the monochromator, and the light beam from the exit slit falls on the cathode of the photomultiplier detector. The resulting a.c. signal is amplified by a variable gain amplifier, and converted to d.c. in a synchronous rectifier circuit for operating a galvanometer. The output signal may also be fed to an external potentiometric recorder. The mains voltage input is fed to a constant voltage transformer the output of which supplies the instrument power supply circuit. The photomultiplier EHT supply is further stabilised. Two shutters are provided, one (S1) being between the lamp and the flame so that the lamp can be cut off for emission work. The other (S2) is between the flame and the mirror system for the monochromator input, and cuts off the light beam for zeroing.

On this single beam instrument, after zeroing and allowing the lamp to warm up, the wavelength round the spectral line of the lamp is scanned and the gain of the amplifier adjusted until a peak signal level of about 90% transmission is obtained. The flame is then lit and the burner adjusted. With atomised de-ionised water, or the solvent being used, the amplifer gain is adjusted for 100% transmission on the meter. The sample is then introduced and the meter reading noted. An automatic sample changer unit is available and is very useful for routine analysis. Thirty-two samples in polythene cups of 4 ml capacity may be accommodated. Using this instrument detection limits of 0.05 ppm for calcium and 0.006 ppm for magnesium are quoted.

## Double beam instruments

Double beam time shared techniques may also be used and the only commercial example is the Perkin Elmer 303 atomic absorption instrument. The emission from the hollow cathode lamp is split into two beams by a sector mirror rotating at 1800 rpm. One beam passes through the flame while the other does not. The two beams are recombined by a semi-transparent mirror and passed through a grating monochromator to the photomultiplier detector. After amplification, the signal is separated into sample and reference channels by a vibrating reed chopper, working in synchronism with the sector mirror. These two channels feed a null balance potentiometer circuit. The operator turns the reference slide wire control until zero is indicated on a meter at the output of the null balance circuit. The amount the potentiometer slide wire (coupled to an absorption counter) is moved, represents the difference voltages between the reference and sample channels, and therefore the absorption of the sample beam. The stability of this double beam system enables detection limits of 0.002 ppm and 0.0003 ppm to be obtained for calcium and magnesium respectively.

In addition to the split beam method, two wavelength optical systems have been devised. This method, described by Menzies (40) and Robinson (50) uses two spectral lines of different wavelength, one of the lines not being absorbed by the sample so it can be used as a reference.

## Resonance detectors

Resonance detectors have now been developed for atomic absorption work (55). Within the detector a cloud of atoms of the metal to be determined is produced, and the emission radiation from the hollow cathode focused on it. The atom cloud absorbs radiation of the resonance line, and subsequently produces characteristic resonance radiation proportional to the absorbed line. The detector thus functions as a monochromator of fixed wavelength with a pass band of the order of 0.002 Å. With the burner/atomiser unit between the lamp and the detector, we have a conventional atomic absorption spectrophotometer. The wavelength and sensitivity stability are however claimed to be superior to a conventional monochromator. This equipment is manufactured by Techtron Ltd. under licence to C.I.S.R.O. in Australia. In the model AR-200 instrument, the light source is modulated at 285 Hz and the output of the photomultiplier is then passed through a synchronous demodulator. Recording and digital readout accessories are available.

# FLUORIMETRY

When a molecule is optically excited, its energy level is raised and light is absorbed. When the molecule returns to the ground state, fluorescent light is emitted a short time (0.01 to 0.001 sec) after absorption, the fluorescence spectrum being a mirror image of the incident light. Light emission occurring after a longer time delay (0.1 m sec to 10 sec), is called phosphorescence. With very dilute material the fluorescence intensity is proportional to concentration for a constant excitation wavelength.

High pressure mercury lamps are usually used in filter photometers for the incident light source, as most substances are easily excited to fluorescence in the ultra-violet region. The resulting fluorescence beam may be 500 or 1,000 Å greater than the wave length of the incident radiation.

## Instrument circuits

A photometer on which fluorescence measurements can be made is the Eppendorf instrument. A primary filter is placed between the mercury lamp and the sample, and a secondary filter between the sample and photomultiplier detector. The secondary filter absorbs at the primary incident light wavelength so preventing scattered primary light being picked up by the photomultiplier. It is usual to position the photomultiplier at such an angle to the incident beam that the detector views only the illuminated sample. Slits are used to prevent reflections from the cell walls falling on the photomultiplier. For precision work a monochromator should be used, and two fluorimeters of advanced design using high intensity xenon

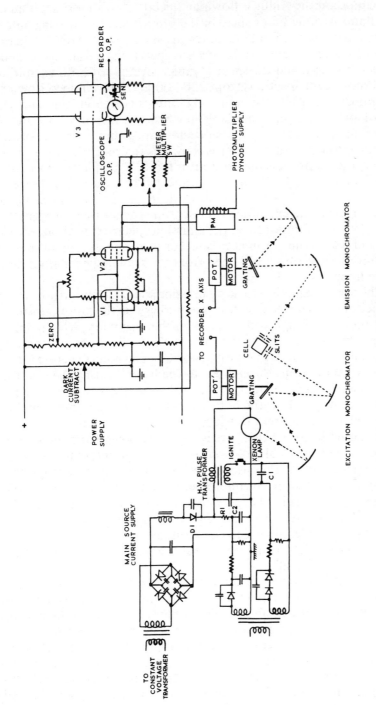

*Fig. 3.17.* The Aminco Bowman Spectrophotofluorimeter.

arc lamps are the Aminco Bowman Spectrophotofluorimeter and the Baird Atomic Fluirispec. In the Fluorispec, two grating monochromators are used in tandem for dispersing the incident radiation, and a further two for the fluorescent spectrum. A 150 W d.c. xenon arc lamp is used and the spectral range covered for both excitation and fluorescent spectra is from 220-700 m$\mu$. The Aminco Bowman instrument (6) (33), also uses a high intensity xenon lamp and a xenon-mercury lamp is available in which the mercury lines are superimposed on the xenon continuous spectrum. A circuit of the Aminco Bowman fluorimeter is shown as figure 3.17. In the photometer, the photomultiplier output is fed to a high impedance differential amplifier V1 V2 which is connected to a cathode follower output circuit V3. With the motor driven emission monochromator scanning, and the excitation monochromator set to a wavelength for maximum fluorescence, the emission spectrum is obtained. If the emission monochromator is set for peak energy, the fluorescence excitation spectrum is obtained from the scanning excitation monochromator.

The xenon lamp is struck by pressing the ignition button, when capacitor C1 is discharged into the primary of the pulse transformer. This produces a high voltage pulse in the secondary, which ionizes the gas in the lamp. Once the gas is ionized, the arc is momentarily sustained by the charge on capacitor C2. The voltage across C2, drives direct current through R1 to the lamp, heating the electrodes and so enabling the main current to flow. As C2 discharges, the voltage at diode D1 cathode falls and this diode can then conduct the main source current flowing.

It should be noted that care is necessary when handling xenon lamps as they are liable to explode due to high internal pressure.

Attachments for fluorescence work can of course be fitted to many commercial spectrophotometers and one particularly good design is due to Sill (53) and is used on the Cary model 14 and 15 spectrophotometers. Special cells with the rear surface silvered on two adjacent sides are used. The beam incident on the cell is reflected back on its own path so that the photodetector see's only the fluorescent light.

A recent instrument introduced by Beckmans is the Ratio Fluorometer. A mercury lamp with a phosphor coated sleeve is used, and mercury wavelengths above 237 m$\mu$ plus peaks due to the phosphors at 310, 360 and 450 m$\mu$ are obtained. The mercury lamp is of special design with a common cathode and two anodes. The lamp cycles in phase with the mains supply frequency so that excitation energy is applied alternately to the sample and reference cells. Primary filters are placed between each side of the lamp and the cells. The fluorescent radiation is focused on to a photomultiplier detector and passes through a secondary filter. The electronic circuit is similar to the DB spectrophotometer except that a ring demodulator circuit is used in place of a vibrator to separate sample and reference signals.

A Perkin Elmer instrument (54) uses a beam splitter so that a portion of the light beam falls on a fluorescent screen monitored by a photomultiplier. The output from this photomultiplier and of the fluorescence detector photomultiplier are fed to a ratio recorder and with the fluorescence of the screen proportional to the intensity of the exciting beam (independent of wavelength) a true excitation spectrum is obtained.

**Phase and polarisation fluorimeters**

A high performance phase fluorimeter has been described by Müller et al., in which the exciting beam is modulated by an electro optic (Pockels effect crystal) modulator at up to 27 MHz. Part of this light is reflected from a quartz plate on to a reference photomultiplier and the remainder excites the sample. Fluorescent light is detected by a second photomultiplier, and the phase difference between the fluorescent and reflected light is measured by a wide band phasemeter circuit. This technique may be used for the study of fluorescent life times (43).

The polarisation of fluorescent light reflects the molecular geometry of the sample solution. A recording fluorescence Polarisation photometer has been described by Deranleau, in which the xenon lamp emission is passed through a monochromator and polarising prism on to the sample cell. The fluorescent light is monitored by two separate photomultipliers each viewing the cell through a polarising prism. This enables the components of the light emission, vibrating in parallel and perpendicular to the direction of propagation of the exciting polarised light, to be measured. The photomultiplier outputs are passed to an analogue computer and various functions are generated from the two signals and plotted on an XY recorder (17).

# POLARIMETRY

In the spectro-polarimeter, optical rotation of the plane of polarisation of linear polarised light is measured as a function of wavelength. A high intensity light source (xenon arc, sodium or mercury vapour lamp) and monochromator are used.

**Recording instruments**

The sample is placed between a polarising and an analyser prism, the planes of polarisation of which are crossed (i.e. at right angles), so that normally extinction occurs. The analyser prism is oscillated at a low frequency around the crossed position and an alternating light signal results from the photomultiplier detector. This is amplified and passed to a phase sensitive detector, where the signal is split into components corresponding to the direction of deflection of the analyser prism. When an optical rotation occurs in the sample cell, these two components are no longer equal and an input signal is fed to a servo amplifier. The servo motor rotates the polarising prism until the zero condition is reached, and this compensating rotation angle may be read off on a counter driven mechanically from the prism drive. The input signal to the servo amplifier is also fed to an XY recorder, the X axis recording wavelengths or time.

**Use of Faraday cells**

Another method of modulating the light beam uses a Faraday cell, which is placed between the servo motor driven polariser and a fixed analyser prism. A Faraday cell consists of a cell containing water, or a cylindrical glass rod inside an electrical coil (21). With an alternating current passed through the coil, the plane of polarisation oscillates at the same frequency as the supply voltage.

This technique is used in the Cary model 60 spectropolarimeter, where the Faraday cell consists of a silica cylinder, surrounded by a coil operated from the mains supply through a resonant filter. The instrument has a double prism monochromator and can automatically record rotary dispersion directly in degrees over a range of 1850 to 6,000 Å. A 500 W xenon source is used. The motor which drives the polariser prism to the null position, also operates a transmitting potentiometer which feeds a follow up recorder servo amplifier.

A Faraday cell is used on the polarmatic 62 spectropolarimeter to compensate for optical rotation, the compensating coil current being measured. This current is proportional to optical rotation, but varies with wavelength, so that a correction factor (Verdet correction) must be applied for each wavelength. The polarmatic 62 spectropolarimeter designed by the National Physical Laboratory is manufactured under licence jointly by Bendix Electronics Ltd and Bellingham and Stanley Ltd. Cultured quartz prisms are used in the monochromator, which in addition to dispersing the beam also act as polariser and analyser elements. The compensator (C) and modulator (M) Faraday cells are situated, as is the sample, in the light path between the two prism elements. The compensator coil has two windings. One is for compensation, and is used in conjunction with a special function potentiometer for Verdet correction, the other being used for injecting a current to give a rotation of 20 millidegrees at 18,300 wave numbers. This facility is useful for assessing the response of the instrument.

A simplified circuit of the polarimeter is shown as figure 3.18. The a.c. signal from the photomultiplier is passed through amplifier and filter circuits, which have an overall pass band centred on the

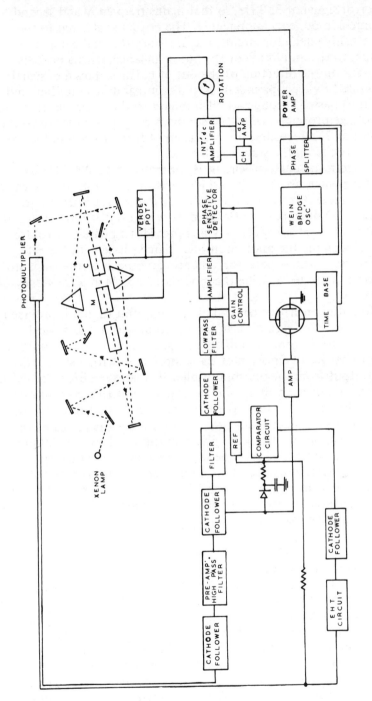

*Fig. 3.18.* The Bellingham and Stanley spectropolarimeter Polarmatic 62.

servo error signal of 380 Hz, so that mains frequency and second harmonic components are rejected. The error signal is fed to the phase sensitive detector circuit, the d.c. output signal being proportional to the amplitude of the out of balance, with a polarity dependent on the direction of the out of balance. This d.c. signal is integrated by a chopper stabilised d.c. integrating amplifier, and the output passed through a milliammeter to the compensating coil C. A voltage proportional to the compensating current is fed to the Y axis of an XY recorder, the X axis input being from a potentiometer driven from the wavelength scale.

The electronic detector circuit is controlled by a Wein bridge oscillator, which uses silicon carbide resistors and polystyrene capacitors for temperature and life stability. A thermistor is used in the oscillator circuit to control the amplitude of oscillations. The oscillator is fed to a phase splitter, which provides a phase control for correctly aligning the photomultiplier signal phase with the reference of the phase sensitive detector, and also an output to the transistorised power amplifier driving the modulator coil M at a frequency of 380 Hz.

Automatic sensitivity control of the photomultiplier is provided to compensate for variations in illumination within the optical system due to absorption, fall off in response of the photomultiplier, or changes in emission of the xenon lamp with wavelength. If the output from the photomultiplier increases, a difference exists between a reference voltage and the signal input to a comparator circuit. This circuit changes the EHT voltage, so that the sensitivity of the photomultiplier is adjusted to counteract the increase, and the a.c. signal output from the photomultiplier then is maintained at its original level. The cathode ray tube enables the quality of light being received by the photomultiplier tube to be assessed, so that conditions of high scattering, absorption or fluorescence will not lead to false results being obtained. A transmission meter is also provided (not shown).

### Measurement of circular dichroism

Circular dichroism is exhibited by optically active samples having unequal absorption coefficient for left and right circularly polarised light. This may be measured using attachments to a spectropolarimeter, and in the Cary model 60, circular dichroism is recorded directly in terms of ellipticity in degrees. Ellipticity is defined as the arc tangent of the ratio of the minor to major axis of the elliptical polarisation. The beam from the monochromator is passed through the polariser prism and electro-optic modulator into the sample cell. If the sample exhibits circular dichroism, left and right circularly polarised light is transmitted unequally. This results in a current from the photomultiplier composed of an alternating component superimposed on an average component. These two components are applied to the measuring circuit, and the pen servo system produces a direct recording of the circular dichroism.

## RAPID SCAN TECHNIQUES

### Instrument circuits

In recent years, a great many new techniques have been used for the detection of transient chemical species with very short life times, and for the study of enzyme kinetic reactions. An instrument for the study of fast reactions and high speed recording of constant spectra has been described by Niesel et al. (45). A selected portion of the spectrum is moved 100 times/sec across the exit slits of the monochromator by a mirror oscillating at 50 Hz. A second mirror oscillating at 12 kHz, passes monochromatic light alternately through the sample and reference. A single photomultiplier is used and its output is fed to a logarithmic amplifier. The 12 kHz oscillator, which controls the mirror, operates a synchronous switch which passes signals from the logarithmic circuit to two integrator units. The integrator output voltage is proportional to the area of the pulses. This voltage is stored for one cycle, and at the end of each cycle the difference between the voltages is determined and fed to the vertical plates of an oscilloscope, the deflection of which is proportional to absorption of light in the sample. The oscilloscope horizontal deflection is proportional to wavelength and is controlled by a voltage in phase with the oscillation of the 50 Hz mirror. In this way, a complete absorption spectrum appears on the screen every 10 milliseconds.

### The storage oscilloscope

Rapid kinetic reactions may be studied with a special version of the Phoenix dual wavelength scanning spectrophotometer already described. This version uses a storage oscilloscope in place of the XY recorder, and events taking place in the range 30 milliseconds to 1 second may be examined. A diagram of a storage oscilloscope tube is shown as figure 3.19. The tube has two electron gun systems, with a storage grid mesh behind the phosphor coated screen. The first gun system (writing gun), is conventional with X and Y deflecting plates, while the second is without deflecting plates and produces a parallel beam of low velocity electrons over the whole screen area (flood gun). Initially the storage mesh is held at a negative potential, which cuts off the low velocity electrons from the screen. High velocity electrons from the writing gun strike the storage mesh and secondary electrons are produced. They are attracted to the positive collector mesh, and the charge distribution stored in the capacitance

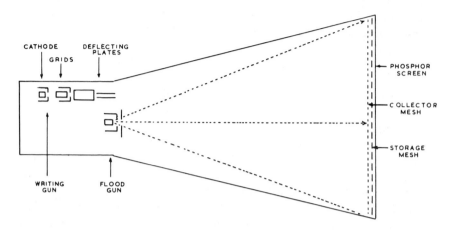

*Fig. 3.19.* Storage oscilloscope tube.

between the storage mesh and the backing electrode. This is not affected by the low velocity electrons. The storage mesh goes positive at the point where secondary electrons were produced, and the low velocity electrons are allowed to pass to the fluorescent screen. The signal may be stored without display by switching off the low velocity electron gun. When display is required, the gun is switched on. Split screen storage tubes are available which use two flood guns. Tektronix storage oscilloscopes use this method, and one half of the screen may be used for storage while the remainder is used for normal display. The entire screen may also be used for storage or display. Collimation bands, round the tube between the guns and the target screen, control the flood gun electrons so that they cover the face of the tube evenly.

### Infra-red systems

Rapid scan techniques are also used on infra-red systems and Herr and Pimentel (32) describe an instrument with fast response photoconductive detectors using which, transient chemical species with life times of 50 microseconds have been detected. Rapid scan infra-red instruments with cathode ray tube presentation of spectra are very useful in analytical chemistry. Mechanical scanning mechanisms for 10-100 second scan times have been described by Daly and Sutherland (14), Powell (47) and Baker & Robb (4). For faster scan times, of the order of 10-100 milliseconds, rotating or oscillating prisms in the optical system are used. A double beam in time spectrometer, having a scan repetition frequency of between 5 and 25 Hz and a chopping frequency of 2,000 Hz, described by Daly (15), uses a nernst glower as the source and a lead sulphide cell as a detector. The detector signal is superimposed on a 2 kHz ripple voltage, and is modulated on to a 200 kHz carrier which is passed through an automatic gain control amplifier. A signal output is obtained, by filtering and rectification, which is proportional to the difference in absorption between the two beams.

Normal infra-red instruments usually take several minutes to scan their spectral range, but for some purposes, such as analysing the effluent from a gas chromatograph or fast reaction studies, a scan time of a few seconds is desirable. An instrument designed for these purposes is the Beckman IR 102 spectrophotometer, which scans from 2.5 to 14.5 microns in 5 or 12.5 seconds. Repetitive scans with six second cycles (1 second between 5 second scans) are possible but for fast reactions, with the wavelength fixed and absorbance recorded against time, components with a life time of 10 milliseconds may be recorded. A polychromatic beam from a nernst glower, is passed through the rotating chopper and sample cell, to one of three wedge filters mounted on a rotating wheel. A limited wavelength band is passed by the filter to the slit mechanism, which is mechanically adjusted as the scan proceeds, so that the beam energy remains relatively constant. The filters are of increasing thickness from the 2.5 u end, so that transmitted wavelengths increase linearly in three ranges to 14.5 u. A 120 HZ triangular wave signal from the thermocouple detector is amplified and applied to a high pass filter, which gives a square wave output. The electronic circuit is transistorised except for the amplifier input stages which use miniature ceramic valves. The square wave is passed to a phase splitter which feeds a full wave demodulator using photodiodes optically controlled by the chopper. The d.c. signal is filtered by a fast response integrator and passed to a high speed recorder.

## STOPPED FLOW SPECTROPHOTOMETERS

**Principle**

In kinetic measurements, two solutions whose reaction is to be determined are admitted to a mixing chamber and then passed through an observation tube, positioned between a monochromatic light source and a phototube. The flow velocity may be registered by a potentiometer coupled to the plungers of the admitting syringes. The potentiometer output is differentiated and amplified, the flow velocity and spectrophotometer traces being recorded against time. In the stopped flow method, data are obtained when the mixed reactants are stationary in the observation tube.

**Instrument circuits**

Chance and others have detailed many methods used for the study of enzyme reactions, in which a sensitivity $1 \times 10^{-4}$ OD is required with low zero drift, in order to allow several minutes of recording at high sensitivity (11). Circuits have been described which use photocell control of the light source intensity with a stabilised lamp voltage supply. A circulating fan to reduce modulation of the light beam, due to rising hot air and dust, was used and with a chopper stabilised amplifier, a signal to noise ratio of 50,000:1 obtained over the wave length range 380-580 m$\mu$. Chance and Legallais used a stopped flow attachment on a Beckman DU monochromator. The small current ($8 \times 10^{-11}$ A) from the photocell was amplified by an electrometer valve then converted to a.c. by a chopper a.c. amplifier, demodulated and the d.c. output used to drive a recorder. It was necessary to stabilise the heater voltage for the hydrogen lamp supply (12).

Enzyme systems are obtained as turbid suspensions and the concentration is such that 90% of the incident light is lost in scattering and absorption. Since absorption bands of components may be separated by only about 10 m$\mu$, two monochromators are used and light from each is alternately passed through the sample and detected by a photomultiplier. The square wave output from the photomultiplier represents the difference of light absorption at the two wavelengths used. For turbid samples, one wavelength is set at the isobestic point for the reaction of interest, and the other used to monitor the reaction.

In the method described above, due to Chance, a 60 Hz vibrating mirror was used for beam switching, and in the Niesel design described earlier a mirror operating at 12 kHz. In order to overcome mechanical problems, a system using two xenon lamps modulated to give square light pulses 180° out of phase with each other at 5 kHz, has been described by DeSa and Gibson (16). A Colpitts type oscillator, with phase shift networks and amplifiers, is used to obtain four 5 kHz signals with appropriate timing, which are used to synchronise the system. Schmitt trigger pulse generators modulate the xenon lamp currents, and the two monochromators are set to wavelengths $\lambda r$ and $\lambda m$. Fibre optics are utilized as a light guide to the reaction chamber. Two photomultipliers are used, and a part of the combined light is passed to tube 1 (monitor), and the remainder to the reaction chamber and tube 2 (measuring). The monitor tube output is independent of the chamber reaction, and any variation is due to lamp intensity changes etc. The measuring photomultiplier is affected by both variations and the chamber reaction.
The photomultiplier signals are fed to a transistorised synchronous chopper which produces 4 outputs. These are $\lambda m\, \lambda r$ from tube 1, and $\lambda m\, \lambda r$ from tube 2. The difference, $\lambda m$ or $\lambda r$ (tube 2) $- \lambda m$ or $\lambda r$ (tube 1), is obtained by using a difference amplifier which gives an output corrected for pulse amplitude variations not due to absorbance changes in the sample. The corrected $\lambda r$ and $\lambda m$ outputs are fed to a dual channel oscilloscope, which displays absorbance changes occurring in the reaction chamber at the two wavelengths

This system can detect optical density changes as small as $1 \times 10^{-3}$ and rapid chemical reactions up to 250 secs$^{-1}$ in turbid suspensions can be observed.

It has been reported that difficulties may arise in some stopped flow methods where colour changes occur in the course of the reaction, and where several chemical species are present whose spectral bands overlap. In order to overcome these problems, Dye and Feldman (19) have applied a rapid scan monochromator to a stopped flow apparatus in which a selected spectral region, between 0.2 and 10 $\mu$, is scanned at rates between 3 and 150 scans per second. Two photomultipliers are employed and the anode currents are fed to a logarithmic circuit, which comprises a high input impedance amplifier, dual logarithmic transconductor, operational amplifier output and a power supply. The output is proportional to the absorbance of the sample and during a run data are stored on magnetic tape. The output of the logarithmic circuit, a synchronising signal pulse, and an audio signal are recorded on separate channels. The recorded information may be played back into a storage oscilloscope.

The Durrum Instrument Co. market a developed Gibson stopped flow spectrophotometer, which has a wavelength range of 350 m$\mu$ to 800 m$\mu$ using a tungsten iodine source. A storage oscilloscope readout system is used and reaction half times as short as 5 milliseconds may be measured.

## FLASHING LIGHT SPECTROPHOTOMETERS

### Use in photosynthesis studies

This type of spectrophotometer may be used for studying fast reactions occurring during photosynthesis. Such an instrument has been described by Bacon, Treharne and McKibben (3), which records the absorption spectrum of reaction intermediates with a monitoring light at a specific time after photo-excitation. For reaction kinetic studies, a series of experiments with a variable time delay are carried out, or the change in absorbance with time followed at a particular wavelength. The constant intensity beam from the grating monochromator is split by a prism into two parts. One is passed to a reference photomultiplier and the other through the sample cuvette to the measuring photomultiplier. The beams are balanced by a neutral density optical wedge. The S-11 response photomultiplier outputs are applied to a difference amplifier the output of which feeds an oscilloscope. A computor for average transients is used to extract the signal from high background noise.

The light flashes are obtained from xenon tubes which are controlled by multivibrator circuits. The output of an astable multivibrator operates a thyratron which supplies a pulse, via a transformer, to trigger the first xenon tube. The second flash tube can be operated at the same time as the first, or after a delay set by a monostable circuit. The flash duration is 20 $\mu$ seconds and transient optical density changes as small as 0.0001 OD can be detected. Other types of flash sources are quartz tubes filled with rare gas and the ruby laser (excitation at 694 m$\mu$). The laser overcomes the problem of afterglow, encountered with ordinary flash tubes, which prevents observations until several milliseconds after the flash (11).

## SIGNAL ENHANCEMENT BY DIGITAL COMPUTER

**Computer for average transients**

Digital computers have been used with many types of spectrometers for signal enhancement. In the method of Cuthrell and Schroeder (applied to a Perkin Elmer spectrophotometer) a potentiometer is driven by the recorder pen drive shaft and a signal proportional to the pen position is fed to the computer. Signal/noise and signal/background ratios can be increased by factors of 600 and 6 respectively in only six two minute scans. This enables noise obscured absorbance peaks to be resolved (13).

The type of computer used in signal enhancement is known as a computer for average transients (CAT). The technique may be used with any high sensitivity spectrometer in such applications as studying the kinetics of enzyme or fast chemical reactions, and nuclear measurements. The principle is that the event of interest is recorded many times and the signal integrated out of the background noise. The signal plus noise may be converted into pulses in a voltage to frequency converter, and applied to a multi channel pulse height analyser with a digital memory. The input pulse rate is proportional to the signal voltage. The analyser channels are opened in succession and each channel corresponds to a definite part of the spectrum. Pulses corresponding to the signal of interest are always fed to the same channels and the stored pulses increased. Pulses corresponding to noise however, are random, and increase only as the square root of the number of scans while the signal increases directly with the number of scans. The ratio of stored signal to noise pulses therefore increases rapidly with the number of scans. The memory store can be addressed through a register so that the enhanced signal can be monitored visually on a display oscilloscope or recorder at any time.

## References

1. Alkemade, C. T. J. (1954) *Doctorial theses.* University of Utrecht.
2. Archibold, E. (1957) *J. Sci. Inst.* **34**, 240.
3. Bacon, Ke., Treharne, R. and McKibben, C. (1964) *Rev. Sci. Inst.* **35**, 3, 296.
4. Baker, E. B. and Robb, C. D. (1943) *Rev. Sci. Inst.* **14**, 362.
5. Berger, L., Peterson, J. and Budge, E. (1965) *Rev. Sci. Inst.* **36**, 93.
6. Bowman, R. L., Caulfield, P. A. and Udenfriend S. (1955) *Science* **122**, 32.
7. Brealey, L. and Ross, R. E. (1951) *Analyst* **76**, 334.
8. Burks, H. (1965) *Rev. Sci. Inst.* **36**, 1375.
9. Cary Application Report (1964) A.R.14-2. Applied Physics Corp.
10. Chance, B. and Legallais, V. (1951) *Rev. Sci. Inst.* **22**, 627.
11. Chance, B., Eisenhardt, R. H., Gibson, Q. H. and Longberg-Holm, K. K. (1964) *Rapid Mixing and Sampling Techniques in Biochemistry.* Academic Press.
12. Chance, B. and Legallais, V. (1951) *Rev. Sci. Inst.* **22**, **619**, 627.
13. Cutherell, R. and Schroeder, C. (1965) *Rev. Sci. Inst.* **36**, 1249.
14. Daly, E. F. and Sutherland, G. B. (1946) *Nature* **157**, 547.
15. Daly, E. F. (1960) P277 in Electronics for Spectroscopists. ed. C. G. Cannon. Hilger and Watts Ltd.
16. DeSa, R. J. and Gibson, Q. (1966) *Rev. Sci. Inst.* **37**, 900.
17. Deranleau, D. A. (1966) *Anal Biochem.* **16**, 438.
18. Dixon, M. (1967) *Biochem. J.* **104**, 585.
19. Dye, J. and Feldman, L. (1966) *Rev. Sci. Inst.* **37**, 154.
20. Estarbrooke, R. (1962) *Anal Biochem.* **3**, 369.
21. Faraday, M. (1846) *Phil. Mag.* **28**, 294.
22. Gilford, S. R. and Wood, W. A. (1961) *Anal Biochem.* **2**, 589.
23. Gilford, S. R. and Wood, W. A. (1961) *Anal Biochem.* **2**, 601.
24. Gillham, E. J. (1956) *J. Sci. Inst.* **33**, 338.
25. Golay, M. J. E. (1947) *Rev. Sci. Inst.* **18**, 357.
26. Gordy, E., Hasenpusch, P. and Sieber, G. F. (1964) *Electronic Engineering* **12**, 808.
27. Hallman, N. and Leppänen, V. (1949) *Suomen Kemistileht.* **22B**. No. 11. 55.
28. Halperin, A. and Braner, A. (1957) *Rev. Sci. Inst.* **28**, 959.
29. Hardy, A. C. (1935) *J. Opt. Soc. America* **25**, 305.
30. Harriss, L. (1946) *J. Opt Soc. America* **36**, 597.
31. Heidel, R. H. and Fassel, V. A. (1951) *Anal Chem.* **23**, 784.
32. Herr, K. and Pimentel, G. (1965) *Appl. Optics.* **4**, 25.
33. Howerton, H. K. (1959) *J. Opt. Soc. America* **6**, 50.
34. Jones, W. G. and Walsh, A. (1960) *Spectrochem. Acta.* **16**, 249.
35. Kaye, W. and Waska, F. (1964) *Anal. Chem.* **36**, 2380.
36. Levikov, S. I. and Shishatskaya L. P. (1961) *Optics and Spectroscopy* December.
37. Lippman, D. I. (1967) *Carygraph.* Vol. 4, No. 3. Applied Physics Corp.
38. Marr, A. and Marcus, L. (1961) *Anal Biochem.* **2**, 576.
39. Massman, H. (1963) *Z. Instrumentenk.* **71**, 225.
40. Menzies, A. C. (1960) *Anal. Chem.* **32**, 898.
41. Michaelson, J. L. and Liebhafsky, H. A. (1936) *Gen. Elec. Rev.* **39**, 445.
42. Mitchel, R. L. (1950) *Spectrochim. Acta* 462.
43. Müller, A., Lamry, R. and Kokubun, H. (1965) *Rev. Sci. Inst.* **36**, 8, 1214.
44. Nielson, S. (1955) *Rev. Sci. Inst.* **26**, 516.
45. Niesel, W., Lübbers, D., Schniewolf, D. and Richter, J. (1964) *Rev. Sci. Inst.* **35**, 5, 578.
46. Peterson, W. C., Bauman, R. P. and Price, I. W. (1966) *Rev. Sci. Inst.* **37**, 10, 1316.
47. Powell, H. (ed) (1950) Symposium on Spectroscopic Methods in Hydrocarbon Research. Inst. of Petroleum.
48. Pritchard, B. S. and Holmwood, W. A. (1955) *J. Opt. Soc. America* **45**, 690.
49. Ramsay, J. A., Falloon, S. W. H. W. and Machin, K. E. (1951) *J. Sci. Inst.* **28**, 75.
50. Robinson, J. W. (1961) *Anal Chem.* **33**, 1226.
51. Sharpe, J. (1962) *Industrial Electronics* **1**, 70.
52. Sharpe, J. (1961) *Electronic Technology* June 1961.
53. Sill, C. (1961) *Anal. Chem.* **33**, 1579.
54. Slavin, W., Mooney, R. W. and Palumba, R. W. (1961) *J. Opt. Soc. America* **51**, 93.
55. Sullivan, J. V. and Walsh, A. (1966) *Spectrochim Acta* **22**, 1843.
56. Sweet, M. H. (1950) *J. Mot. Pict. Tel. Engrs.* **54**, 36.

57. Thompson, A. E. (1966) *Spectrovision* **15**, 7, Unicam Ltd.
58. Trüjillo, E. F. The Model DKA Ratio Recording Spectrophotometer, Beckman Instruments.
59. Vallee, B. L. (1954) *Nature* **174**, 1050.
60. Walsh, A. (1952) *J. Opt. Soc. America* **42**, 94.
61. Walsh, A. (1953) *J. Opt. Soc. America* **43**, 58.
62. Walsh, A. (1955) *Spectrochim Acta* **7**, 108.
63. Walsh, A. (1962) L.S.U. Intern. Symposium Modern Methods of Analytical Chemistry Jan. 1962.
64. Warren, R. L. (1952) *J. Sci. Inst.* **29**, 284.
65. Wormser, E. M. (1953) *J. Opt. Soc. America.* **43**, 15.
66. Zak, B., Holland, J. and Williams, L. (1962) *Clin. Chem.* **8**, 530.

# 4 Spectrometry

So far we have described ultra-violet, visible and infra-red absorption spectrophotometers. This chapter is concerned with various other types of spectrometers in which lines or peaks characteristic of the sample under investigation are measured. The wavelengths involved ranged from the x-ray region to radio waves. Gamma spectrometers are described with other nuclear equipment in chapter 5. The characteristics of various analytical spectrometer systems are outlined in tables 4.1 and 4.2.

## EMISSION SPECTROSCOPY

In emission spectrometry, radiation characteristic of the element is emitted when the sample is vaporised by means of a flame, arc, or spark. The radiation emitted is dispersed through a prism or grating on to a photographic plate or a photoelectric detector. Flame methods have already been discussed and by comparison excite only a few spectral lines.

### d.c. excitation methods

In the d.c. arc, voltages between 50 and 300 volts are usually used, the circuit comprising a rectifier unit supplying the arc current through a reactance in series with the electrodes. The reactor reduces the tendency for the arc to wander and flicker.

Elements below the limit of detection of the spark method can be detected with the d.c. arc, which has a good line to background ratio. If carbon electrodes are used in air, cyanogen molecules are formed, which emit band spectra in the region of 3,600 to 4,200 Å and so may hide emission from elements such as molybdenum, iron and the rare earths. This may be overcome if a Stallwood jet is used, in which a curtain of gas (argon-oxygen) is maintained around the sample electrode. Cyanogen bands are eliminated and excitation efficiency is increased, as the sample burns slowly and is cooled by the gas (53). Vacuum spectrographs are used for elements such as carbon and sulphur.

| System | Flame | Emission spectrograph arc or spark discharge | Atomic absorption | X-ray fluorescence |
|---|---|---|---|---|
| Principle | Light emitted by atoms excited by flame. | Light emitted by atoms excited by electrical discharge. | Chemically unbound atoms of sample vapourised in a flame, absorb radiation from a hollow cathode lamp at the resonance line. | Sample excited by X-rays. Intensity of X-ray fluorescence radiation measured in quantitative analysis (wave-length gives qualitative information). |
| Characteristics | Small sample volume. Invariably a solution. Lower flame temperatures than emission spectrograph. (1800-3100°.) | Only small sample volume (few mg) required for analysis. Usually solid. Intensity of spectral lines for one element affected by presence of other elements. High discharge temperatures. (Spark—10,000°K, arc—6,000°K.) | Metallic and semi-metallic elements in solution or suspension can be determined. Hollow cathode lamps for wide variety of elements available. Rapid analysis (few minutes). | Elements from sodium onwards can be analysed. Sample not destroyed and can be solid or sometimes liquid. Rapid method. Subject to absorption effects (radiation emitted by elements from sodium to potassium absorbed by air–vacuum used). |
| Performance | Sensitivity for sodium about 0.01 mg/L. Much lower for other metals. | High sensitivity, particularly for metals (e.g. sodium 0.1 μg/ml–spark). | Standard deviation on same samples over long period 1-2%. Very high sensitivity (e.g. 0.005 μg/ml potassium and sodium; 0.0003 μg/ml magnesium; 0.002 μg/ml zinc). | Good reproducibility ($2\delta = 0.2\%$). High sensitivity (few μg or p.p.m. detected). Generally superior to emission spectrographs at high concentrations. |
| Instrumentation | Flame photometers use filters. Flame spectrophotometer uses quartz monochromator. Essential when high flame temperatures used, as in determination of heavy metals. | Permanent record obtained with instruments using photographic plate. Direct reading instruments may be linked to computer for inter element correction and automatic operation. | Monochromator and photomultiplier detector used. Most systems single beam. Modern instruments have digital concentration accessory. | Automatic instruments may incorporate computer for background and inter element corrections. |
| Examples | Most spectrophotometers have atomiser–burner attachments to replace normal light source, e.g. U.V. and visible instruments of Zeiss, Beckman, Optica. | Hilger & Watts polyvac system. | Unicam SP 90, 900. Perkin Elmer 290, 303. | Hilger & Watts X-ray spectrometer. Philips X-ray analysers. |

TABLE 4.1.

CHARACTERISTICS OF VARIOUS EMISSION SPECTROMETERS AND SPECTROGRAPHS

# EMISSION SPECTROSCOPY

| System | Mass Spectrometry | Combined Mass Spectrometer – Gas Chromatography | Electron Spin Resonance | Nuclear Magnetic Resonance | Raman Spectrometry |
|---|---|---|---|---|---|
| Principle | Ions are accelerated into a magnetic field and separated according to their masses. | Effluent from chromatograph column passed through molecular separator (removing carrier gas) to ion source chamber of mass spectrometer. | Resonance interaction between a microwave field and electrons in sample, situated in a magnetic field. | Resonance interaction between a radio frequency field and protons in sample, situated in a magnetic field. | Raman wavelength shift observed when light passed through transparent medium containing molecules which undergo a change in polarizability. |
| Characteristics | Rapid analysis of gases, liquids and solids. Sample must be volatile even if only to a low degree. Greater resolution obtained with double focusing systems in which deflecting magnet system followed by an electrostatic sector. | As for mass spectrometer. Both gas chromatograph and solid inlet systems provided. Required sample amount may be less than 0.01 μg for capillary column and 0.1 μg for packed column. | Sample placed in resonant cavity. Rotating, optical transmission & liquid helium cavities available. | Sample placed in resonant cavity. | Monochromatic light from source incident on sample tube. Emission from raman tube passed through monochromator and detected by photomultiplier. Liquid samples to about 10 μL. Solid samples as small as 100 μg. |
| Performance | Range from a mass range of 1-400 with a resolution of 300 (10% valley) on small instruments to a mass range up to 3000 with a resolution of 1000 (10% valley) on larger double focussing instruments. | Less than $3 \times 10^{-9}$ g/sec of methyl stearate may produce a detectable parent mass peak. | Resolution of order $2/10^5$. | Resolution may be better than 1 part in $10^8$. | Resolution may be 1.5 cm$^{-1}$ Normal frequency range of 0-5000 cm$^{-1}$ frequency shift from Toronto lamp 4358 Å exciting line and about 30-4000 cm$^{-1}$ frequency shift with 6328 Å HeNe gas laser. |
| Instrumentation | Sector magnets usually 60, 90 or 180°, and may be permanent or electromagnet. Field strengths up to 14 k gauss. Ion accelerating volts variable from about 700 to 3500. Electron beam energy 5-100 V. | As for mass spectrometer. Infra-red spectrophotometer used in interrupted elution technique. | Requires stable and homogeneous magnetic field. Microwave klystron and detection system may operate at 9.6 GHz | Requires very highly stable and homogeneous magnetic field. System may operate at 60 MHz or 100 MHz. (Typically). | Grating monochromator(s) used. Output from photomultiplier, amplified and displayed on chart recorder. Gas lasers only recently introduced. |
| Examples | AEI-GEC MS 10. Perkin Elmer 270. Hitachi-Perkin Elmer RMU-6E. Consolidated dynamics. | LKB 9000 Perkin Elmer 270 GC-DF. | Varian E.3 | Varian A60. HA-100 Perkin Elmer R10, R12. | Perkin Elmer LR-1 Cary 81. |

TABLE 4.2.
CHARACTERISTICS OF ANALYTICAL SPECTROMETER SYSTEMS

### a.c. excitation methods

High voltages of 1,000 volts or more are used in a.c. arcs, which are steadier and more reproducible than d.c. arcs. The arc voltage is obtained from the secondary of a transformer, and care must be taken as the voltage used is lethal. The arc voltage, current, and electrode separation must be carefully controlled for reproduceable results, and a thyratron controlled arc has been described by Badocz (2).

The a.c. spark method uses a high voltage transformer (up to about 50 kV), and the electrodes, in parallel with a capacitor, are connected across the secondary winding. An inductance in series with the electrodes reduces the excitation of lines and bands from air molecules. The current on initiation of the spark is given by $V\sqrt{(C/L)}$, where V equals applied potential, C equals capacitance, and L equals inductance. The characteristic of the spark depends on the value of L and C, and if the inductance is increased, the spark characteristic becomes similar to that of the a.c. arc. Feussner has described a circuit which uses an auxiliary rotating spark gap driven by a synchronous motor. The rotating gap is closed for a short time at the peak of each half cycle, and the number of the decay cycles of the spark (which are displayed oscillographically) are controlled, giving reproduceable excitation (20).

### Lasers and gas plasmas

An optical ruby laser beam can be used to vaporise samples, and one example is the Jarrell-Ash Co., microprobe, in which laser excited vapour short circuits the gap between two electrodes.

A radio frequency excited gas plasma torch has recently been developed and may be used for spectrographic analysis. A plasma is obtained by utilising the high conductivity of an ionised gas. High frequency current (2 kW at 34-38 MHz) passes through an inductor, which encompasses the gas, so that energy is transferred, giving rise to temperature in the heart of the plasma of 10,000-20,000 K. No electrodes are used, and a very low background is claimed over the range 2,300 to 5,000 Å. Sensitivity is of the same order as that of the d.c. arc in air, although the plasma has a stability comparable to a chemical flame. Equipment incorporating a plasma torch made of silica is available from Radyne Ltd and can be used with most types of spectrographs and flame spectrophotometers. Using a high frequency plasma on a Unicam SP900 spectrophotometer, concentrations of elements found in p.p.m. (1% of minimum concentration to give full scale deflection) are, Calcium 0.005, Magnesium 0.03 and Zinc 4. More recently experiments have been carried out using a 25 kw generator operating at 6 MHz (46).

Spectrographic electrodes may be composed of the materials being investigated, or if the material will not stand high temperatures, and is not a conductor, the sample is placed in a hollow of the carbon or graphite cathode electrode.

## Spectrographs

In qualitative analysis, a comparison may be made between the spectrum of known components alongside the spectrum of the sample, or the sample spectrum compared with standard spectra. In quantitative analysis, the intensities of spectral lines are compared. With a spectrograph employing photographic plate techniques, developing equipment and a densitometer are necessary. Since we are concerned with electronic instruments, only the direct reading spectrograph will be described, which as photomultiplier detectors are used, eliminates the need for the above photographic apparatus.

Direct reading instruments have a polychromator in which a series of slits are used to select the spectral lines of interest. The light passing through each slit is focused on to a photomultiplier tube, a series of tubes being used to measure specific spectral lines. The output from each photomultiplier charges capacitors, and the resulting voltage is a function of the concentration of the element. Accurate analysis cannot however be simply performed by measuring these voltages directly, and various methods of measurement using comparison circuits are described below.

One simple method, is to precharge a comparison capacitor which is then discharged to the level of the charge on the analyser capacitors. The time taken for the charge on the comparison circuit to equal the analyser capacitor charge is measured. This time is related to the percent concentration of the element.

The capacitors are charged during the spark exposure period, the comparison circuit being charged from the output of a photomultiplier energised by the line from an internal standard. This is selected so that its spectrum does not interfere with the sample lines of interest, and is incorporated with the material under investigation in the electrode. The voltages on the analyser capacitors, are in effect, then measured as a voltage ratio relative to the internal standard.

In another method, the charge on each analyser capacitor is compared with a standard capacitor charged by a circuit including a 10 kHz crystal oscillator. The time taken to raise the charge on the comparison capacitor, until it equals the charge on the analyser capacitors, is measured and displayed on a dekatron counter.

A recent development by Kessel and Jecht uses a rotating electrode holder in which the sample and a standard are sparked alternately. A synchronising switch directs signals alternately to either the standard capacitor or the analyser capacitor. At the end of the spark period the difference between the stored charges is measured (33).

Correction for the background can be made automatically, and in the method of Weekley and Norris, the background emission is measured by a photocell, and the signals proportional to background, are electronically subtracted from the line plus background signals stored on the elements capacitors (57).

## Automatic and computerised spectrographs

In direct reading spectrometers, the capacitor integrates the collected current from the photomultiplier. If a high impedance valve volt meter is used to measure the capacitor potential, the readings obtained must be referred to a calibration curve in order to determine the concentration of the element. This is complicated by the fact that the reading for one element may be dependent on the concentrations of other elements in the sample. Smith and Rippon have, however, applied analogue computing techniques to this problem, and use a function generator to produce a voltage output directly proportional to concentration (52). The function generator, connected between the photomultiplier and voltmeter, divides the calibration curve into a number of linear segments by means of

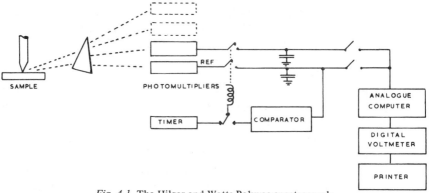

*Fig. 4.1.* The Hilger and Watts Polyvae spectrograph.

switched resistor networks. The error due to the multisegment approximation is claimed to be less than 0.25% in low alloy steel analysis. Nine resistance bridge circuits are paralleled at the summing junction of the amplifiers in the function generator in order to simulate a complete curve. A low leakage polystyrene capacitor is used as a storage device to enable an element concentration to be corrected for the presence of other elements in the sample.

Automatic spectrographs have been developed by Hilger and Watts Ltd (figure 4.1) in which voltages, representing the concentrations of the elements in the sample, are stored on the capacitors. Several sets of matched capacitors are used, so that the results from say three sparkings can be routed to different capacitor stores, and an average obtained by a store selection and an averaging sequence circuit. A digital voltmeter measures the stored voltages and its output is fed to a printer. Provision is made for print out of the results of individual sparkings, the average of a number of sparkings and sample identification data (21).

Teleprinter links between the laboratory spectrograph and various departments of a steel works are being increasingly used. Digital computers may be utilised to accept inputs from the spectrograph capacitor stores and carry out linearity and inter-element corrections automatically. Several digital computer systems have been described (49) (37).

In the Hilger and Watts system, there are up to 25 photomultipliers and the sample of steel to be analysed is sparked in an argon atmosphere. The discharge occurs between the prepared face of the sample and a counter-electrode of either silver or tungsten. After a pre-integration period allowing conditions to stabilise, the output from each photomultiplier is fed to a separate capacitor. The voltages on these capacitors are measured sequentially at the end of the integration period and printed out on an electric typewriter. For low alloy steels, the concentrations corresponding to these readings are then read off calibration graphs for each element. For high alloy steels, the capacitor voltages are fed to an analogue computer which carries out all the necessary calculations, including corrections for inter-element effects, and gives a print-out directly in concentration, summing to 100%. This is completed within 1½ minutes of mounting the sample on the discharge stand.

## RAMAN SPECTROMETRY

**Principle**

Raman observed in 1928 a scattering of light by molecules of compounds, the wavelength of the scattered light having a different frequency to that of the incident radiation. The Raman effect is expressed in terms of the Raman shift, that is, the difference in wave numbers between the incident light and the emitted Raman line. The effect is observed when monochromatic light is passed through a transparent medium containing molecules which undergo a change in polarisability. The frequencies of the Raman lines are characteristic of the molecular vibrational mode and the intensity of the lines is useful in quantitative analysis. Raman lines appear on either side of the line of incident radiation, the strong lower frequency line being termed the Stokes line and the weaker upper frequency the anti-Stokes line.

**Light sources**

The emitted Raman lines have perhaps only 1/100 of the intensity of the incident light and high intensity light sources are required. Banks of four to eight H type mercury arc lamps may surround the sample tube and are enclosed in a hollow reflective jacket. The sample in the Raman tube must be protected from the intense lamp heat, and this is achieved by surrounding the tube with a glass water jacket through which cold water is circulated. Liquid filters, used to obtain monochromatic light, are placed between the water jacket and sample tube. The 4538 Å mercury line is commonly used and the Torronto lamp has a good ratio between this line and unwanted lines at 4338 and 4349 Å. The Torronto lamp is a popular source and consists of a low pressure water cooled spiral mercury lamp, the sample tube being placed in the centre of the spiral (32).

Recently continuous gas lasers have been used as sources (35) (36). Commercial instruments using Helium-Neon Lasers are manufactured by Perkin Elmer and Cary. Emission is at 6328 Å. The Raman emission is passed through a monochromator and focused on to a photographic plate or a phototube. With photographic methods of detection several hours exposure time is required and with gases up to a days exposure may be necessary.

**The Cary 81**

An example of a Raman spectrophotometer utilising electronic techniques is the Cary model 81. This instrument is available with both a Torronto lamp or a 50 mW HeNe laser. Emission from the Raman tube is focused on to an image slicer, which converts the rectangular apertures at the tube into a series of thin slices, which are then re-assembled lengthwise to correspond to the monochromator entrance slit. In this way, the maximum amount of Raman emission is passed to the double grating monochromator. Light from the monochromator exit slits is chopped at 30 Hz by a rotating mirror, so that radiation falls alternately on two S20 response photomultipliers. The output pulses from the photomultiplier tubes are of opposite phase but a special circuit arrangement, in which the final dynode of one photomultiplier is connected to the anode of the other, combines the signals in phase. (With the He/Ne laser a single end on photomultiplier is used.) A portion of the exciting radiation also passes through the rotating mirrors to a reference photomultiplier. After amplification, the Raman signal and the reference signal are compared in a null balance recording circuit. The output from the reference amplifier feeds the recorder slide wire, the slider for which is connected to the Raman signal amplifier input. The signal amplifier output drives the pen motor and a tachogenerator gives feedback stabilisation.

## MASS SPECTROMETRY

**Principle**

In Mass spectrometry, an incident electron beam is used to remove electrons from atoms of a sample placed in an ionisation chamber. The resulting positive ions are then accelerated by a high potential into a magnetic field. The magnetic field is at right angles to the direction of motion of the ions, so that these particles move on a circular path. With the accelerating voltage and field strength constant all particles of the same mass follow the same radius trajectory, and are focused on to a collector which is usually connected to an electrometer. The resulting signal is further amplified and recorded. If the accelerating voltage or field strength are varied (within limits), ions of different masses are focused on to the collector, so that a series of peaks known as the mass spectrum are obtained. The position of the peaks indicates the mass of the ion and the peak height is proportional to the number of ions.

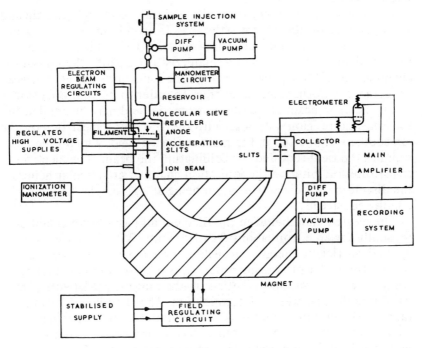

*Fig. 4.2.* Mass spectrometer principle.

Mass spectrometers consist basically of five units (1) the ion source (2) the sample inlet system (3) the electrostatic accelerating and magnetic system (4) the ion collector and (5) the amplifier and recording system. A 180° magnetic deflection system mass spectrometer is shown in figure 4.2. A very low pressure of the order of $10^{-7}$ mmHg is maintained inside the deflection tube to prevent the ion stream colliding with gas molecules. The sample must also be introduced into the ionisation chamber without the pressure in the chamber rising above about $10^{-5}$ mmHg. Two separate vacuum systems are usually used each with a rotary vacuum pump, diffusion pump, pressure measuring and safety devices.

## Sample inlet systems

Various techniques for the introduction of samples have been developed. These include break off devices and the introduction of liquid through a micro-pipette/sintered disc arrangement under mercury. Gases may diffuse through a molecular leak (such as a minute hole in metal foil) into the ion source. Solid samples can be ionised by evaporation from a thin heated filament or by the use of a radio frequency spark for solids with a high melting point. Heated inlet systems are used for liquids and solids with low volatility.

**The ion source**

The ion source chamber is usually maintained at $10^{-5}$ mmHg and a temperature of 200°C. An orthodox electron bombardment source is the Nier type, in which electrons emitted from an incandescent ribbon filament pass through a small positive slit into the ionisation chamber (40). After passage through the ion source the electrons are collected at a positive anode or trap. In order to prevent diffusion of the electron beam an auxiliary magnetic field may be applied to the ionisation region, although with the 180° type spectrometer the ion source is immersed in the main analysing field, which then supplies the field required for collimating the beam. The electron trajectories on entering the ionisation chamber are circular helices, the direction of the electron beam being at right angles to the incoming gas stream.

The electric field between the filament and ionisation chamber is usually variable from about 6 to 100 volts. With 50-70 volts applied, sufficient energy is available to dissociate the molecules, so that this order of ionising potential is used to obtain a mass spectrum. Lower ionisation potentials, in the range say 9-14 volts, are used in molecular weight determination. The flat plateau region (about 50 volts) of the ionisation potential v ionisation efficiency curve is chosen as a working range. Variation in the electric field will then have a minimum effect on peak intensities. The positive ions formed, are accelerated towards negatively charged slits by a positive repeller plate at the entrance to the ion source. The final accelerator slit may be charged at the start of a scan to a potential of 4,000 volts which is then allowed to leak away at a controlled rate.

**Ion collection**

After passing through the analyser tube, placed in the field of a 180° or sector magnet, the ion beam passes through slits on to a collector. Measurement errors due to the loss of secondary emission electrons from the collector may be prevented by the negative charge on additional repeller slits, as these electrons are then returned to the collector. A suppressor slit, positioned between the collector and repeller slits, may be used to control the admission of ions to the collector, and therefore can be used to control the effective

# MASS SPECTROMETRY

width or resolution of the system. It is however, necessary to place an earthed slit between the collector and suppressor to screen the collector against potential disturbances.

### Amplifying and recording systems

The collector is usually a Faraday cylinder or cage although electron multipliers are also used. The ion beam current is very small ($10^{-10}$ to $10^{-18}$ amperes), and consequently an electrometer amplifier stage followed by a high gain feedback amplifier is necessary. A vibrating reed electrometer amplifier followed by an a.c. amplifier may also be used (30). It is possible to overcome the problem of d.c. amplification by modulating at a low frequency the potential of the ionisation repeller plate or the output of the electron gun filament (19).

A typical pre-amplifier unit for a mass spectrometer might have an input impedance of $10^{10}$ ohms with a response time of 0.5 secs. Pre-amplifiers with rise times of 0.06 secs may be used on fast scan instruments. The main amplifier gain may be variable over a range of 1,000:1.

In the electron multiplier, the incident particles pass through a transparent grid onto a tungsten cathode. Glass strips with a high resistance coating are used to form field and dynode strips. External magnets evenly spaced along the length of the multiplier provide a uniform magnetic field perpendicular to the electric field. The electrons emitted from the cathode therefore follow a partly cycloid motion along the multiplier. Secondary electrons are produced from the dynode surfaces and are collected at the anode, a gain of 10 electrons per incident ion being obtained. The gain may be varied by changing the negative high voltage applied to the dynode. In the 14 dynode multiplier used on the LKB9000A mass spectrometer, a change of voltage from 1.2 kV to 3.5 kV varies the gain from $10^3$ to $10^7$. The E.H.T. supply is stabilised so that a short term stability better than 0.01% is obtained over a 10 minute period.

The peaks which will be recorded may vary by a factor of 1 to 1,000 in amplitude, so that if strip chart recorders are used, an automatic range changing switch is incorporated (9). Multiple trace recording oscillographs are also used, and may consist of five separate galvanometers each of a different sensitivity. The galvanometers record simultaneously and the reading is made from the most sensitive trace remaining on scale.

### Regulating circuits

Electronic regulating units are used to control (1) the emission from the source filaments, (2) the current supply to the deflecting electromagnet, and (3) the voltages applied to the various spectrometer slits. To control the filament emission, the collected electron beam is passed through a resistance, and the resulting voltage developed is applied to an amplifier whose output controls the filament supply voltage. Many emission regulators have used the principle of the Ridenour and Lampson design in which the amplifier controls the load on a transformer secondary winding, an auxiliary secondary winding supplying the filament. If emission falls, the transformer load is reduced and the secondary winding voltage increases so as to counteract the decrease in emission. Regulation to ±½% is possible (48).

It is also necessary to control the temperature of the source region to accuracy of about ±0.1 C. Temperature changes following filament current control may lead to variations in the electron and ion beams, and circuits have been described which maintain constant power input to the filament. The anode trap current is used to control the voltage applied to the first slit of the electron accelerating gun, this slit having an effect on the distribution and energy of the electron beam (10).

Many circuits have also been described for controlling the magnet current, and on commercial instruments stability is better than ±0.01%. Scanning by the electrostatic method, in which a capacitor in the accelerator slit circuit of the electron gun is charged, and then allowed to discharge through a resistance network, has been referred to earlier. Magnetic scanning by supplying the magnet current from a motor driven potentiometer circuit is also common.

There are many types of mass spectrometer and the operation of various instruments is described below.

### (1) Magnetic Deflection Spectrometer

The spectrometers described previously are of this type, in which the ions are segregated into beams each of a different m/e ratio (m = mass, e = charge). Deflection through 180° or a sector of 60°, 90° or 120° are the methods used. The magnetic fields must be uniform and electro-magnets are used on large instruments with fields variable up to about 13 kilogauss. Smaller instruments may use alnico permanent magnets with field strengths about 4-6 kilogauss.

### (2) Isotope Ratio Spectrometer

Two ion beams are produced and passed through a double exit slit to the collectors. The two signals are amplified simultaneously, and the larger of the two is attenuated by resistors until a null balance is obtained.

### (3) Double Focusing Spectrometer

An electro-static analyser is placed between the accelerating slits and the magnetic field. This arrangement focuses ions having the same m/e ratio but different initial velocities and directions. Narrower slits may be used and an increased resolving power obtained.

### (4) Cycloid Focusing Spectrometer

A uniform electrostatic field is applied at right angles to the magnetic field and in line with the accelerating ions, so that they travel cycloid paths instead of a circular path. Ions of different energy but of the same m/e ratio are focused on to the collector.

### (5) Bennet Radio Frequency Spectrometer

Ions pass through a number of stages, each comprising three grids, to a potential energy selector and the collector. The central grid of each stage is supplied with a voltage at radio frequency, the outer two grids being at earth potential. With a certain radio frequency voltage, only ions of a particular m/e ratio can pass through the grids and emerge from this analyser section with maximum energy. Heavier ions are rejected, while lighter ions pass through the grid quickly, and so do not attain maximum energy. The potential energy selector deflects the ions having less than the maximum energy away from the collector. The resolution of this type of instrument is not as high as the spectrometers described above.

### (6) Time-of-Flight Spectrometer

The electron beam in the ion source is energised for a very short time period at regular intervals, and the resulting ions are electrostatically accelerated into an evacuated tube. The time taken for the ions to drift through the field free tube to the electron multiplier collector depends on the ion m/e ratio. The collector output is amplified, and displayed on an oscilloscope, the time base of which is synchronised with the energisation of the electron beam. Since the lightest ions reach the multiplier before heavier ions, a complete mass spectrum may be obtained in microseconds. This type of spectrometer is useful for kinetic studies and for analysing the effluent of gas chromatographs.

### (7) Combined Gas Chromatograph-mass Spectrometer

Combined gas chromatograph-mass spectrometer instruments are now available. In the LKB system, shown in figure 4.3 the effluent from the temperature controlled gas chromatograph column passes through a Becker-Ryhage type molecular separator (which removes 95% of the carrier gas whilst 50% to 75% of the sample remains) into the ion source chamber. Before the ions from the source reach the deflecting magnet, they pass through a hole in an electrode plate which extracts a small part of the beam. The output from the electrode plate is amplified and recorded so that a measure of the total ion current is obtained. This depends on the sample leaving the column, providing the column temperature and ion source are constant. A 60° sector magnet and electron multiplier, followed by an electrometer and a wide band amplifier, are used for obtaining mass spectra which are recorded by three galvanometers in the recording U.V. oscillograph.

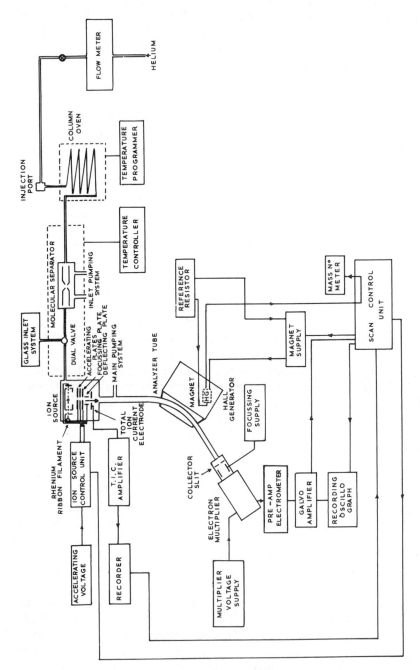

Fig. 4.3. The L.K.B. combined mass spectrometer–gas chromatograph.

A mass spectrum by itself does not always give sufficient information to allow positive identification of chromatograph column eluents. Identification can however be made more certain by using a complimentary recording infra-red spectrophotometer. Eluent from the column is led to the spectrophotometer cell which has a capillary leak to the mass spectrometer. The elution from the column is interrupted, and the infra-red spectrum of a peak scanned before the next peak is eluted. The recorder of the stop-start chromatography system incorporates a peak sensing circuit, which initiates the sequence of trapping components, recording spectra etc., as the peak emerges.

The interrupted elution gas chromatograph technique was pioneered by Dr. R. P. W. Scott of the Colworth House Research Laboratories of Unilever Ltd. Commercial equipment of this type has been described for use with the A.E.I. MS 10 mass spectrometer, and consists of an interrupted elution chromatograph based on the Philips PV 4000, and a Unicam SP 200 GA spectrophotometer with a scale expansion unit and slave recorder.

### (8) Omegatron

The omegatron is a type of small cyclotron in which the electron beam passes through a small aperture into the ionising chamber. The beam is parallel with an external magnetic field and at right angles to a radio frequency electric field applied between two plates, one at the top and the other at the bottom of the chamber. Ions produced gain energy from the radio frequency field and spiral outwards when the field frequency equals the cyclotron frequency of the ion. As the omegatron construction does not allow the use of an electron multiplier collector, its sensitivity is limited.

### Digital techniques

Digitisation of mass spectra can be obtained by computer techniques or by causing the galvanometer beam to deflect across a grating so that electrical pulses are produced (13). Multichannel analysers may be used with high sensitivity mass spectrometers, and a time-amplitude converter for adapting the output of a time of flight spectrometer to a multichannel analyser has been described by White. A sawtooth oscillator is used, and a gating circuit (operated by shaped and amplified ion pulses from the spectrometer) samples a ramp voltage so that the gate output is a pulse whose voltage level is dependent on the time between the start and stop signals. These pulses are fed to the multichannel analyser and a histogram of pulse heights is obtained (58).

# RADIO AND MICROWAVE SPECTROSCOPY

## Nuclear magnetic and electron spin resonance

The nuclei of atoms may be considered as minute magnets spinning about an axis. Normally they point in all directions so that their total field is zero. If however, the nuclei are placed in a strong homogeneous magnetic field, every nucleus assumes a definite orientation with respect to the field; each orientation of the nucleus corresponding to a different energy level. With a radio frequency electric field applied at right angles to the magnetic field, energy is absorbed by the nucleus if the alternations of the field corresponds exactly to the frequency of the nucleus energy level. The absorption of energy causes the nuclei to change from one energy level to the next highest, the radio frequency absorption spectrum containing lines which give information on the chemical nature of the nucleus. The spectrum may be scanned by varying the oscillator radio frequency or by slowly changing the magnetic field. The latter method is usually used in nuclear magnetic resonance (NMR) spectrometers, the change in the applied field changing the spacing of the energy levels.

Atoms that have an odd number of electrons also exhibit magnetic properties, the electrons being able to interact with a magnetic field. If such chemical species are placed in a constant magnetic field and subjected to a microwave frequency, transitions occur at resonance and the absorbed microwave energy causes the electron spins to change from one energy level to the next highest. Instruments operating in the microwave region are known as electron spin resonance (ESR) or electron paramagnetic resonance (EPR) spectrometers. In nuclear magnetic resonance we are principly interested in resonance lines of protons and a few other nuclei such as fluorine.

## Radio frequency spectroscopy

Two groups of research workers in the United States originally developed techniques for observing nuclear magnetic resonance. Purcell Torrey and Pound observed proton resonance, the sample being placed in a resonator circuit. The output from this circuit was amplitude and phase balanced against another tuned circuit, the bridge arrangement being fed from a 30 MHz oscillator. The magnetic field was changed slowly, and when absorption occurred, the Q of the resonator circuit decreased. The bridge then became unbalanced and an output signal was obtained on a microammeter of the detecting system (45).

Block, Hansen and Packard used a different technique in which a separate coil, placed at right angles to both the d.c. magnetic field and the radio frequency applied field, was utilised to detect absorption. In this crossed coil method, no voltage is induced in the detector coil until the magnetic field is adjusted to the point at which magnetic resonance occurs. At this point, magnetisation is greatly affected by the radio frequency field and components of magnetisation exist at right angles to the applied field, causing a voltage to be suddenly induced in the coil (4).

A Block diagram of a crossed coil NMR spectrometer is shown as figure 4.4(a). In practice it is not possible to eliminate electrical coupling between the input and the detector coils, and in order to overcome this difficulty, a small degree of coupling is mechanically introduced to compensate for the existing coupling. The mechanical arrangement consists of an adjustable semicircular copper paddle which can be moved across the end of the detector coil. A signal generator supplies power to the Y coil and the X detector coil out-

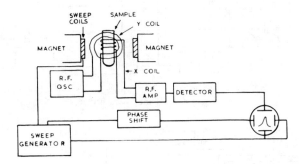

*Fig. 4.4.* (a) Basic NMR Spectrometer.

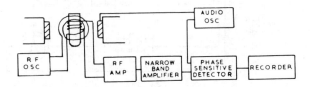

*Fig. 4.4.* (b) High sensitivity field modulation.

put is fed to a radio frequency amplifier, which is tuned to the frequency of the signal generator. The Y transmitter coil is in fact wound in two halves to enable the sample to be inserted. Coils are also wound around the pole pieces of the magnet to allow a sweep or modulation of the magnetic field. This permits oscilloscope display of sample absorption lines, the oscilloscope time base being fed (via a phase shift circuit) from the sweep or audio signal generator. The output from the radio frequency amplifier is rectified and passed through an audio frequency amplifier to the Y plates of the oscilloscope.

Modulation of the field with an applied sinusoidal wave obtained from the audio frequency oscillator and the use of a phase sensitive detector gives high sensitivity. The method is shown in figure 4.4(b). In order to obtain a display corresponding to sample absorption, it is necessary to measure only the component of the induced voltage in the X coil that is in phase with the voltage applied at the Y coil. This is achieved by the use of the phase sensitive detector, the output of which may be displayed on a chart recorder. The recorder trace is proportional to the first derivative of the line shape. Instead of modulating the field, the same effect can be obtained by frequency modulating the radio frequency oscillator. Typical modulation frequencies used are in the range 25 to 400 Hz.

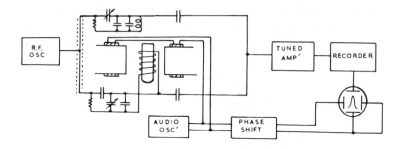

*Fig. 4.4.* (c) Tuned R.F. bridge.

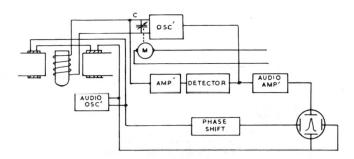

*Fig. 4.4.* (d) Feedback oscillator.

Field modulation is not only applied to the crossed coil system, but also to the resonance absorption spectrometer, and a diagram of a modulated radio frequency bridge is shown in figure 4.4(c). This technique was used by Bloembergen, Purcell and Pound (6).

Another type of circuit uses the sample coil as part of the tuned circuit of an oscillator, and when magnetic resonance occurs the Q of this circuit is lowered. The resultant change in radio frequency is amplified, detected and then displayed on an oscilloscope as shown in figure 4.4(d). The Pound Knight Watkins spectrometer is of this type, except that the signal at the modulation frequency is fed to a phase sensitive detector through a narrow band width amplifier, and the output of the phase sensitive detector recorded. The capacitor C is driven by synchronous motor when searching for resonances (42) (43). The spectral noise of this type of circuit has been examined by Howling (28).

Since a line width of only 1 Hz may be obtained in high resolution systems, the audio frequency detection systems described earlier cannot be used and heterodyning circuits are utilised. A simplified block diagram of such an instrument is shown as figure 4.4(e). This technique is used in the Varian 4311 spectrometer, and a 60 MHz radio frequency oscillator is utilised. The output from the sample coil is amplified and passed to a mixer stage, where the 60 MHz signal is heterodyned with a 55 MHz wave from a crystal controlled master oscillator. The resulting 5 MHz wave form is then fed to a phase sensitive detector, the reference voltage for which is obtained via a phase shift circuit after mixing the 60 MHz oscillator output with the 55 MHz from the crystal controlled master oscillator.

A considerable number of circuits have been described in the literature and a revue of some of these which use autodyne and super regenerative receivers has been given by Gutowsky et al (25) (41) (47).

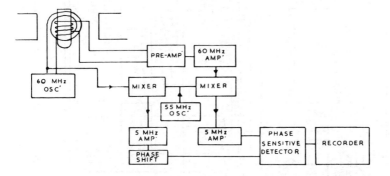

*Fig. 4.4.* (e) Superheterodyne detection.

## Magnet control circuits

In high resolution spectrometers, the field strength must be homogeneous to 1 part in $10^8$ for accurate interpretation of spectra. This requires large diameter magnet faces which are highly polished. Small coils may also be incorporated on the poles and are used in conjunction with a field stabilising circuit. Balancing currents are passed through these coils in order to optimise field homogeneity (22) (38). In another technique described by Block, field inhomogeneities are averaged out by spinning the sample about the vertical axis (5).

Highly stabilised power supply circuits are required for the electro magnets and even if permanent magnets are used precise control of temperature is necessary. Conventional magnets cannot be used above about 23,000 gauss in NMR spectrometers, although superconducting magnets with fields up to 50,000 gauss and field

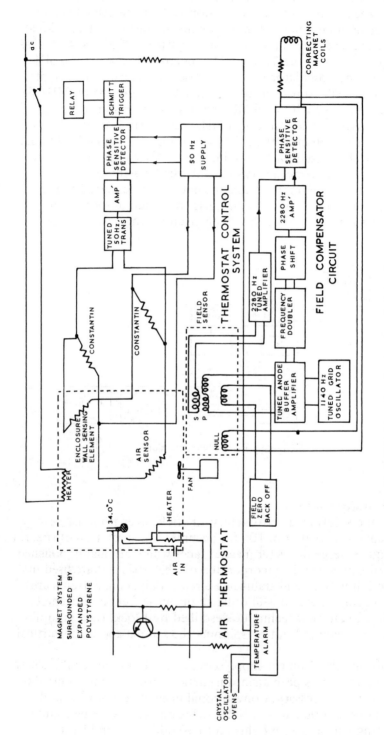

Fig. 4.5. N.M.R. Magnet environment control circuits.

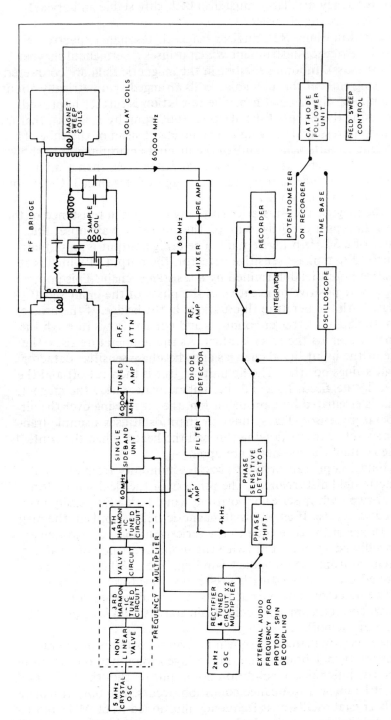

*Fig. 4.6.* The Perkin Elmer R10 N.M.R. spectrometer.

homogeneouses of 1 part in $10^6$ are now being developed. The radio frequency oscillator must also be highly stable and crystal controlled circuits are used.

The Perkin Elmer R10 nuclear magnetic resonance spectrometer is a high resolution instrument which utilises a permanent magnet (14,000 gauss). Inhomogeneities in the magnetic field are countered by passing current through golay coils arranged symmetrically about the sample in the magnet gap. The resolution is further improved by spinning the sample tube at up to 3000 rpm. by means of thermostated compressed air driving a turbine at the head of the tube. The electronic circuits which control the magnet environment are shown in figure 4.5. These circuits give a field drift of less than 1 milligauss per hour under a temperature change of less than 1°C per hour (between 15°C and 28°C).

In the magnet thermostat circuit, any changes in resistance of the copper air sensor or bifilar wire on the enclosure wall, upset the balance of a 50 Hz bridge circuit. The bridge output is amplified and passed to a phase sensitive detector, the output of which is at a positive potential determined by a resistance chain. This point swings more positive or negative as the phase of the input signal changes with temperature fluctuations in the bridge. If the bridge cools, the Schmitt trigger conducts and current flows through the magnet heater via the relay contact. As the temperature rises, the phase of the input signal changes and the phase sensitive detector output swings positive. The Schmitt trigger is then cut off and the heater de-energised. To avoid thermal gradients inside the magnet, the air is circulated by a propeller fan, the air flowing over the air temperature sensor. The spinner air thermostat uses a simple transistor switch, current flowing through the heater when the contacts of the contact thermometer are open.

A field compensator circuit provides immunity from external magnetic field disturbances. The primary of the field sensor consists of two coils in series opposition, wound on a thin easily magnetised wire core. If an external magnetic field is applied, the magnetic fluxes in the core (due to the series windings) are no longer balanced, and an output at twice the input frequency of 1140 Hz is obtained from the secondary winding. The 2280 Hz signal is amplified and compared with a reference waveform in a phase sensitive detector. The output from this circuit passes current through correcting magnet coils. Gain stabilisation is achieved by the use of the nulling winding.

The measuring circuit is shown in figure 4.6 and is of the radio frequency twin T bridge type. The bridge is mounted on the probe, and in the null balance condition no output reaches the R.F. amplifier until a nuclear resonance condition occurs. The output from a 5 MHz crystal oscillator is frequency multiplied to 60 MHz and fed to a single sideband unit which produces a 60.004 MHz output. This is obtained by modulating the 60 MHz by a 4 kHz waveform and using a lower sideband and carrier suppressor circuit to remove the unwanted components. The 60.004 MHz output from the R.F. bridge at resonance, and a 60.000 MHz signal are mixed to give a 4 kHz output. This is further amplified and fed to a phase sensitive detector, the output of which is passed to either a recorder or oscilloscope. The N.M.R. spectrum is scanned by varying the current through the sweep coils, using a signal derived from either the recorder chart potentiometer or the oscilloscope time base. The single sideband system provides good stability of spectrum baseline and of the phase of N.M.R. signal. In addition this system avoids the occurrence of a series of resonances, which would be found if the centre band and lower sideband were not suppressed.

**Recent N.M.R. developments**

In recent years, the technique of using a feedback loop between the magnetic field and the applied radio frequency has been developed. In this method, the radio frequency is automatically adjusted to compensate for any changes in the magnetic field so that the resonance condition is maintained for a standard absorption line.

Stability requirements are not so stringent in broad band or wide line spectrometers, and in fact, E.S.R. instruments may be converted to wide line N.M.R. spectrometers. The field strength is of the order of 1,700 to 3,500 gauss with a homogeneity to 1 part in $10^5$.

Recent developments in N.M.R. spectrometers, include the application of pulse techniques and signal enhancement by repetitive scanning. Torrey applied pulse techniques to a tuned radio frequency bridge resonance absorption spectrometer by modulating the output of the radio frequency oscillator using a gating generator. Square pulses at the nuclear resonance frequency result, and are amplified, detected and passed to the Y plates of an oscilloscope. The X deflection system is triggered by the pulse generator. This method is useful due to its fast response and can be used to measure relaxation times (55). This is also true for another method devised by Hahn, in which the signal is measured after the pulse application. If two short radio frequency pulses are applied, the detected wave form shows two pulses, the trailing edge of the first pulse decaying exponentially (determined by the spin-spin relaxation time and field inhomogeneities). A short time after the leading edge of the second pulse, an echo pulse is observed due to the nuclei emitting radiation. The amplitude of this echo pulse is used in calculations to determine relaxation times (26).

**Microwave spectroscopy**

There are three main groups of microwave spectroscopy (1) Gaseous

(2) Paramagnetic resonance absorption and (3) electron resonance in metals and semiconductors, measurements on free radicals etc.

The early gaseous spectrometers used a klystron to produce microwave radiation, which was fed, using a wave guide, to the absorption cell. This consisted of a cavity resonator and the output from the cavity was detected by a crystal diode connected to a galvanometer. A wave meter was also coupled to the wave guide to monitor the output frequency and power of the klystron (3) (23).

Absorption in the microwave region, was however first observed by Cleeton and Williams, who used split anode magnetrons as the radiation source (11). At present, klystrons are almost always used for producing microwaves. Magnetrons give high power and are more difficult to tune than klystrons, which are used in microwave spectroscopy as only low power is required in order to avoid saturation effects in the sample. The klystron power is usually less than 40 mW, the wave lengths covered being typically within the ranges 3.1-3.5 cms and 0.6-0.71 cms. In paramagnetic resonance, fixed frequencies in the X (3 cms) K (1.25 cms) or Q (8 mm) wave bands are usually used.

Detection systems used in microwave spectroscopy can be broadly classified as follows. (1) The a.c. modulation method, in which either the frequency of the klystron or the magnetic field is modulated. In this method, when the klystron frequency is close to a spectral line, the modulating frequency will sweep backwards and forwards across the line so that the resulting signal represents the shape of the absorption line. In gaseous spectroscopy, the klystron is modulated, while for paramagnetic studies the magnetic field is usually varied. This is done by a wire loop around the sample, or a coil around the resonant cavity walls. (2) The heterodyne method, in which the output of a local klystron oscillator is fed into a balanced bridge wave guide element, so that an intermediate frequency (30-45 MHz) is produced when absorption occurs.

**The Stark effect**

In practice the heterodyne technique may be combined with source or Stark modulation. In the Stark effect, splitting and shifting of spectral lines occur when the sample is subjected to an electrostatic field. Hughes and Wilson applied a 6 kHz square wave to the electrode in a Stark absorption cell, which consists basically of a length of wave guide closed at both ends by vacuum proof mica windows. The electrode is a brass strip held in position by insulating strips of teflon (29). The alternating electric field from the Stark electrode modulates the detected microwave power, which is passed to an amplifier tuned to the modulating frequency. The klystron is also modulated by a sweep generator, which in addition, feeds the X axis of the display oscilloscope. The absorption signal is fed to the Y plates.

## High sensitivity circuits

Source modulation followed by a phase sensitive (lock-in) amplifier is used in order to obtain extremely high sensitivity. The principle of both the superheterodyne and lock in amplfier detection systems are shown in figures 4.7(a) and (b). It must be appreciated however, that actual spectrometer circuits are in fact rather more complex. The heterodyne circuit shown, is applicable to gaseous

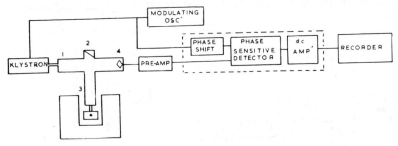

(a) LOCK IN AMPLIFIER SPECTROMETER

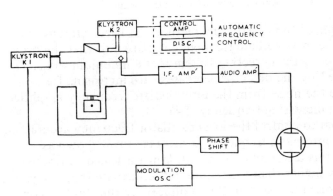

(b) SUPERHETERODYNE MICROWAVE SPECTROMETER

*Fig. 4.7.*

microwave spectroscopy, where the magnetic field is constant and the klystron frequency varied. In paramagnetic spectrometers, heterodyning is simpler as the klystron frequency is constant. In this case, the X plates of the oscilloscope may be supplied with a 50 Hz wave form via a phase shifting network, as in the circuit of England and Schneider (15).

Both diagrams show a balanced bridge system where the bridge may be a magic T or hybrid ring wave guide element. In these devices, the microwave power from the klystron source is fed in through arm 1 and is equally divided between arms 2 and 3. No signal is received by the crystal detector at the end of arm 4, unless the bridge is unbalanced by the sample (connected to arm 3) absorbing energy. When absorption occurs, a signal is reflected to the diode detector and an output signal obtained. It will be appreciated that this device will only operate satisfactorily if the bridge elements are correctly matched. In practice, constructional matching is never perfect and the slight asymmetry is cancelled out by turning screws or a tapered graphite sliding "match" in arm 2. The absorption cell in gaseous microwave instruments is a long length of wave guide, while for paramagnetic measurement, resonant cavities are used.

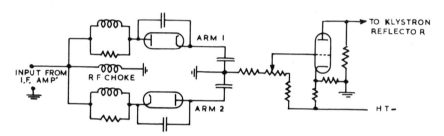

*Fig. 4.8.* Automatic frequency control circuit.

The principle of the heterodyne method has been referred to earilier, and the intermediate frequency is chosen so that the overall noise figure for the detecting system is at a minimum value. The noise generated by the detector decreases with increasing frequency, while the noise from the intermediate frequency amplifier increases with increasing frequency (54).

It is essential to control the local oscillator frequency accurately, and a typical automatic frequency control circuit is shown as figure 4.8. In this discriminator circuit, arm 1 is tuned to resonate at a frequency higher than the required intermediate frequency, and arm 2 at a frequency less than this. Voltages are thus built up on the output capacitors, the polarity and magnitude of the d.c. voltages depending on the variation from the correct value of the intermediate frequency. The discriminator d.c. output, is applied to a control valve, which changes the potential on the reflector of the local oscillator klystron. It will be noticed in figures 4.7(a) and (b) that the oscilloscope Y plates are fed with the modulated frequency via a phase shifting circuit. This is so that the absorbance line traced out on both the backward and forward modulation sweeps coincide.

In the lock-in amplifier method (figure 4.7(a)), the signal to noise ratio is improved due to the phase sensitive detection system. The absorption signal is at the modulation frequency, its amplitude slowly changing as the magnetic field strength or the klystron frequency is slowly changed. The reference input of the phase sensitive rectifier section, is obtained from the modulation source, via a phase shifting circuit. The amplifier output is a d.c. voltage plus a low frequency voltage corresponding to the absorption line.

So far we have ignored the effect of dispersion, which is of course related to absorption as with conventional optical systems. In high frequency spectrometers, the detection system is based on coupling between electrical components and absorption may be considered as introducing a resistance into circuit, while dispersion will cause a change in the effective inductance or capacitance. We can therefore, separate the two components which are both generally contained in the signal from the detector, by use of the phase shifter in the reference input to the phase sensitive detector. For small signals, the absorption and dispersion components are 90° out of phase. The output from the phase sensitive detector is fed through a d.c. amplifier to a pen recorder which traces the differential of the absorption line. The amplitude of modulation is usually small, and the mean klystron frequency (or magnetic field) is swept slowly through the resonance point.

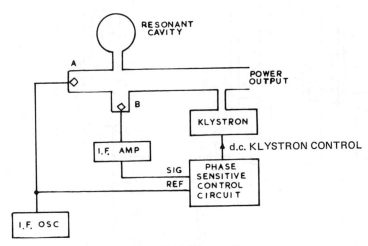

Fig. 4.9. Klystron stabiliser circuit.

## The Pound stabiliser

It is important in the lock-in amplifier detection method, that the absolute frequency of the klystron oscillator is constant while the slow sweep through the absorption line takes place. A commonly used circuit for locking a klystron frequency has been described by Pound, and a simplified diagram of this type of stabiliser circuit is shown in figure 4.9 (44). The intermediate frequency oscillator feeds crystal A with a 30 MHz signal, the resonant cavity being tuned to the klystron frequency. If the klystron frequency changes, the cavity is no longer resonant with it and microwaves are reflected onto crystal B. These waves are then mixed with direct

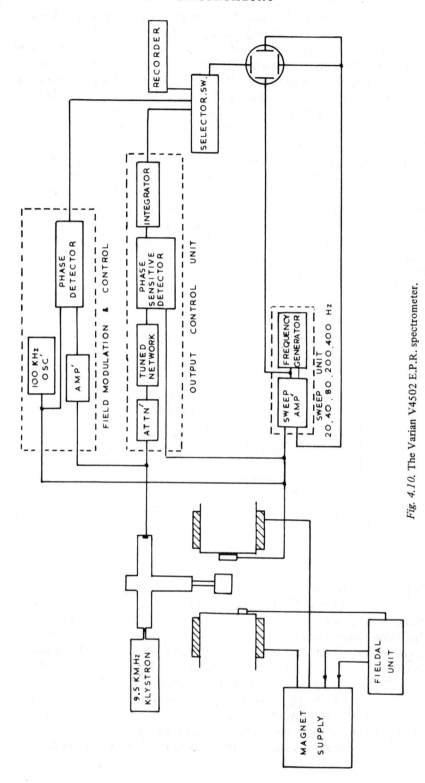

*Fig. 4.10.* The Varian V4502 E.P.R. spectrometer.

microwave power received by crystal B to produce an intermediate frequency, which is amplified and fed to the phase sensitive control circuit. The d.c. output from this unit, changes the voltage on the klystron reflector, so as to restore the frequency to the correct value. The klystron may also be locked to the resonant frequency of the sample cavity instead of to a crystal controlled oscillator or frequency standard. This is done by frequency modulating the klystron at several hundred kilocycles per second. If the klystron and cavity are exactly in tune, no signal at this frequency is received by the crystal detector in the bridge. If the frequency changes, a signal is received, the phase of which indicates the direction of drift. The signal is applied to a phase sensitive control circuit as before.

**The Varian 4502 EPR spectrometer**

The Varian 4502 system is an example of a modern EPR spectrometer, a block diagram of which is shown as figure 4.10. The output from a klystron oscillator, operating at 9,500 MHz (9.5 GHz), and the microwave field are applied to a resonant cavity containing the sample. An X band microwave bridge system is used and the crystal output is filtered, amplified, and phase sensitive detected, both oscilloscope and recorder displays being available. A set of coils on the sample cavity walls are used for modulation. Two separate output control units are provided which are used for audio frequency modulation studies, and 100 kHz field modulation respectively. Audio modulation is useful in initial setting up and for dual sample work. The output (20, 40, 80, 200 or 400 Hz) from a frequency generator is amplified and applied to the modulation coils. An output is also taken from the generator to supply the reference for the phase detection circuit. The crystal detector output is attenuated and passed through a band pass filter network (tuned to the audio frequency used) to the phase detector circuit. In this unit, second harmonics are attenuated and an integrator removes any remaining noise. In field modulation, the output from a crystal controlled 100 kHz oscillator is fed to the modulation coils. The detection system uses a high gain amplifier and phase detector. The sensitivity of the spectrometer when used in this mode is some 15 times that obtainable with 400 Hz audio modulation.

The outputs of both field and audio control units are fed to a selector circuit, which enables either to be connected to the recorder or the oscilloscope. The reflex klystron is controlled by an automatic frequency controlled circuit (accurate to 1 part in $10^6$) and the deviation of klystron and cavity frequencies, the resonator current, the crystal detector current and klystron reflector voltage are monitored by four panel meters. A magnetic field regulating circuit controls the field strength to within 5 parts in $10^7$ for a ±10% line voltage or load impedance change.

Iris holes are used for coupling to the cavity resonator, and a range of adjustments from 3,500 to 2,250 is provided for the multi-purpose cavity. If the dual sample cavity is used, a calibrated standard and the unknown are subjected to different frequency modulation, and the detector crystal output passed through the audio and 100 kHz detection systems described above. The resulting signals are in this case compared on a dual channel recorder. Rotating, optical transmission and liquid helium cavity chambers are also available. A superheterodyne accessory may be fitted and is used with audio frequency modulation, the local oscillator klystron being tuned 30 MHz away from the main klystron frequency.

### Recent EPR developments

So far all the spectrometers described have used microwave bridges, but Faulkner has recently used microwave circulators in an EPR spectrometer which overcomes the disadvantage that half the klystron power is lost in the matching arm of a bridge (16) (17). Basically, the circulator may be considered as a circular wave guide, around which the microwaves can only travel in one direction. Magnetised ferrite inserts rotate the plane of polarisation of the microwaves. Three input-output ports are provided, and if power is fed in from the klystron at point A, microwaves travel round the circulator to port B, which is connected to the resonant cavity. On absorption, power re-enters the circulator and travels to port C, which is connected to the crystal detector by a wave guide. Faulkners spectrometer in fact uses two circulators and a homodyne detector circuit. The all solid state Varian E-3 spectrometer is a commercial instrument utilising a circulator.

Recent developments in microwave spectrometry include the application of rapid recording techniques and signal enhancement by the use of a computer for average transients (34). This type of apparatus is very useful in reaction kinetics, and flow systems have been devised in which two solutions are mixed in a chamber which feeds a reaction tube inserted in the cavity of a spectrometer. This method has been applied to the study of enzyme reactions. The sudden freezing technique in which samples are taken at intervals and deep frozen may also be used (8).

## X-RAY SPECTROMETRY

### X-ray tubes

The X-ray tube consists basically of an evacuated tube containing a heated cathode and a target anode. A potential of about 50 kV is applied to the target, and the electrons emitted from the cathode are accelerated by the high voltage field and quickly brought to rest on impact with the anode. The electrons transfer part of their kinetic energy to the atoms of the target and X-rays are emitted. A continuous 'Bremsstrahlung' spectrum is produced, the intensity rising to a peak then falling off with increasing wave length. The minimum wave length (cut off point) in Å is given by 12,400/V where V is the tube voltage. If the incident electrons have sufficient energy, an electron may be removed from an inner K orbit of the target. This allows an electron from the L or M orbits to drop down to fill the vacancy, so causing the emission of X-ray lines. These lines are characteristic of the target material and are superimposed on the continuous spectrum. Use is made of X-ray fluorescence, as this line emission is called, in the X-ray fluorescence spectrometer, in which a sample subjected to high energy X-rays, emits characteristic lines. Most of the energy of the incident electrons, is however, converted into heat which is removed by cooling with water. The target may also be rotated. The X-rays pass out of the tube through a beryllium or special glass window.

The tube current may be monitored, and a control circuit used in order to vary the filament current so that a constant tube current is maintained. This method (similar in principle to that used in the mass spectrometer) does not work well with an indirectly heated oxide coated cathode due to thermal inertia. These circuits use a controlled impedance, such as a transformer with a variable load on the secondary, in series with the primary of the filament transformer. Circuits have been described using triode valves, thyratrons and transistors (39) (1) (12). Thyratron valves are a source of noise and a recent circuit, due to Yee and Deslattes, uses transistors as the controlled impedance load. The drive to the loading transistors is obtained from a differential amplifier, the input to which is obtained by the difference between the voltage developed across an emission current range resistance and a reference voltage (60). Experiments have also been carried out with control grids in the X-ray tube in order to control the tube current (14) (59).

The voltage applied to an X-ray tube, may be half or full wave rectified using high voltage diodes (kenotrons), and smoothing networks used to obtain a more constant potential. Constant potential voltage, increases the emission of lines from the specimen, particularly at the shorter wave lengths. High energy tubes up to 130 kV are available (7).

## X-ray spectrometers

A typical X-ray spectrometer is shown in figure 4.11 and comprises the X-ray source, the sample, the analysing crystal, beam collimators and the detector system. The source has been described above and the X-rays are emitted in all directions. The sample is placed in a holder and subjected to X radiation, as a result of which, a spectrum comprising characteristic lines with a continuous background is emitted. Collimators are placed between the sample and analyser crystal, and also between the crystal and detector. The collimators consists of a collection of thin parallel plates a few centimeters long, and rays not parallel with the plates are absorbed. The incident radiation on the crystal is at various wave lengths which

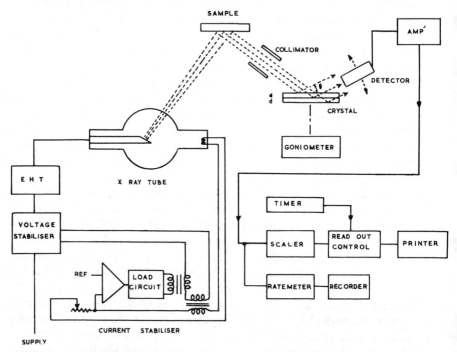

Fig. 4.11. X-ray fluorescence spectrometer.

are separated by reflection from the planes of the crystal lattice. For every angle of incidence, monochromatic X radiation is reflected in accordance with the Bragg formula $n = 2d\sin\theta$, where n denotes the order of the spectrum, d the distance between the reflecting planes of the crystal, and $\theta$ the angle of incidence (and of reflection).

The detector may be a geiger counter, a crystal scintillation counter or a gas flow counter. The latter two are proportional counters, in which the pulse height is directly proportional to the energy of the initiating quantum and inversely proportional to wave length. The pulses are fed to a single channel pulse height discriminator. These devices are described in the next chapter so that only the detector response to X radiation will be considered here.

The argon filled geiger tube with a beryllium window has a quantum efficiency of about 60% over the range 1.5 to 2.1 Å. By comparison, the proportional gas flow counter has about the same sensitivity, but may be used for long wave X-rays as its range extends to around 10Å. An argon methane mixture is usually used with a thin cellulose nitrate, polypropylene, beryllium or teflon window. Other gases such as P-10, methane, helium and neon have been used in the ultra soft X-ray region, in order to obtain better resolution or to lower the anode potential (27) (31) (18). The crystal scintillator has a high quantum efficiency (95%) over the range 0.3 to 3Å, but cannot be used for long wave length X radiation, as this is absorbed in the protective coating of the hygroscopic crystal. In recent years, solid state detectors have been developed which may have 20 times the resolving power of a thalium activated sodium iodide crystal (51). The solid state detector has however a high noise level at low energies and its use is restricted to high energy X-rays (24).

**Automatic X-ray analysis**

For automatic qualitative analysis, the angle $\theta$ between the crystal surface and the incident X-rays is gradually increased, and the intensity of the emitted lines is recorded as a series of peaks. The angular position of the detector in $2\theta$ degree units is also recorded. The detector is geared to the crystal in a goniometer assembly.

In quantitative analysis, the intensity of an emitted line is measured with the detector at the $2\theta$ angle of the peak. The detector pulses are counted until a preset number is reached, or a preset time has elapsed. The goniometer is then set to measure background counts, which are subtracted from the peak counts, the remainder being related to the concentration of the element. Automatic apparatus is available commercially, and several designs have also been reported in the literature. In these instruments, the X-ray tube volts and current, the crystal angle, the detector angle, the pulse amplitude, the amplifier setting, the number of counts and the counting time are all programmed variables. Automatic sample changers and computer links are also incorporated. Wired programme or peg boards are used to set up the analyser and in systems with an automatic sample changer, the goniometer drives to the first angle and all samples are measured at this Bragg angle in turn. When the scaler counts and time have been printed out for all samples, the goniometer moves on to the next programmed angle. This sequence is repeated until the analysis is complete.

Philips have developed an automatic simultaneous X-ray spectrometer (type PW 1250) which is designed for fast routine simultaneous analysis of up to 14 elements in one sample. A maximum of 160 samples can be handled automatically on the sample changer unit. Three modes of measurement are possible (1) absolute, in which the sample is measured for a fixed time, (2) monitor, where the specimen is measured until a pre-set number of counts have been recorded on a chosen channel and (3) ratio, in which the sample is measured for the time required by a previously measured standard to accumulate a fixed number of counts on the same element. The system uses up to 14 separate measuring channels, each with its own collimator, crystal, detector, amplifier, discriminator and scaler. A common timer unit is used. A data processor performs the arithmetical function C = mK + b, where C represents the concentration, I the measured intensity, and m and b are constants representing the slope of the line and background intercept respectively. Correction to m and b can be performed automatically, taking into account interelement effects and background correction.

In methods described for improving the line/background ratio and eliminating interfering radiation, the pulse amplitude discriminator is coupled with the $2\theta$ drive of the spectrometer. In one method, the inverse relationship of the pulse voltage to $dSIN\theta$ is used, the pulses being amplified by a factor proportional to $dSIN\theta$. This is achieved by moving the slider of a sinusoidal potentiometer at a rate proportional to the angular velocity of the crystal. Other methods reported use linear potentiometer function generators or an optical follower which follows a curve of discriminator base line voltage against $2\theta$, an electro mechanical link driving a helical potentiometer controlling the base line voltage as a function of $2\theta$ (50) (56).

**Other X-ray techniques**

The spectrometer illustrated in figure 4.11 uses a flat crystal, and the disadvantage of this method is the attenuation of the incident beam due to absorption in the collimator. This may be overcome by using a curved crystal, its inner surface being ground to the radius of a focusing circle. The geometry of this system however, requires a more complicated mechanism than does the flat plane crystal.

The absorption of X-rays is made use of in the X-ray absorptiometer, in which one half of the X-ray beam is alternately interrupted by a motor driven chopper. The X-rays are then passed through the sample or reference cells and fall on a fluorescent screen, its light emission being detected by a photomultiplier. The photomultiplier output is fed to a phase sensitive amplifier, the reference signal for which is provided by an electro mechanical pulse generator driven by the chopping assembly. An aluminium wedge attenuator is inserted into the reference beam until a balance is indicated on the null meter at the output of the phase sensitive amplifier. The attenuator calibration is a quantitive indication of the composition of the sample.

A radioactive source emitting X-rays can be used to excite a sample which emits secondary X-rays. This radiation is then back scattered on to a crystal scintillation detector feeding a pulse height analyser/scaler unit. This principle is used on the Panax X-ray analyser and typical Bremsstrahlung sources used are 3H/Ti, 147 Pm/Al, 85 Kr/C, and 90 Sr.

**Photoelectron spectrometers**

Information on molecular electron binding energies are obtained in photoelectron spectrometers by bombarding the sample with high energy photons. The photons may be produced by an electrical discharge in helium and a beam of photons is directed through a capillary tube into a chamber containing the vaporised sample. Photoelectrons are produced by collision of the photons with sample molecules and the photoelectrons pass through a slit into an electrostatic deflection analyser consisting of a pair of curved plates with high voltage applied. By varying the voltage applied to the plates, electrons of a particular energy pass through exit slits on to an electron multiplier. The pulses from the multiplier are amplified and fed to an electronic unit containing sweep controls, a ratemeter and recording and display facilities.

### References

1. Allenden, D. (1964) *Phys. Rev.* **133**, A390.
2. Bardocz, A. (1956) *Spectro. chim. acta* **8**, 152.
3. Bleaney, B. and Penrose, R. P. (1946) *Nature* **157**, 339.
4. Block, F., Hansen, W. W. and Packard, M. (1946) *Phys. Rev.* **69**, 127.
5. Block, F. (1954) *Phys. Rev.* **94**, 496.
6. Bloembergen, N., Purcell, E. M. and Pound, M. (1948) *Phys. Rev.* **73**, 679.
7. Blohkin, V. A., Mamonov, Y. M., Belkina, G. L., Sotnikov, V. A., Ovchorenko, E. Y. and Myagkor, G. L. (1965) *Zavodsk Lab.* **31**, 517.
8. Bray, R. C. (1961) *Biochem. J.* **81**, 187.
9. Bunt, E. A. (1951) *Rev. Sci. Inst.* **22**, 58.
10. Caldecourt, V. J. (1951) *Rev. Sci. Inst.* **22**, 58.
11. Cleeton, C. E. and Williams, N. H. (1934) *Phys. Rev.* **45**, 234.
12. Deslattes, R. D. (1964) *Phys. Rev.* **133**, A390.
13. Dudenbostel, B. F. and Klass, P. J. (1959) Advances in Mass Spectrometry (ed.) J. D. Waldron. Pergamon.
14. Eisenstein, A. (1942) *Rev. Sci. Inst.* **13**, 208.
15. England, T. S. and Schneider, E. E. (1950) *Nature* **166**, 437.
16. Faulkner, E. A. (1962) *J. Sci. Inst.* **39**, 135.
17. Faulkner, E. A. (1964) *Lab. Prac.* **13**, 1065.
18. Fischer, D. W. and Baun, W. L. (1964) *Advan. X-ray Anal.* **7**, 489.
19. Forester, A. T. and Whalley, W. B. (1946) *Rev. Sci. Inst.* **17**, 549.
20. Fuessner, O. (1921) *Arc. Eisenhutteniv.* **6**, 551.
21. Glover, B. W., Orwell, R. D. and Adams, P. J. Hilger Journal, Vol. X. No. 1, 7.
22. Golay, M. J. E. (1958) *Rev. Sci. Inst.* **29**, 313.
23. Good, W. E. (1946) *Phys. Rev.* **45**, 234.
24. Goulding, F. S. (1964) *Nucleonics* **22**, 54.
25. Gutowsky, H. S., Meyer, L. H. and McClure, R. E. (1953) *Rev. Sci. Inst.* **24**, 644.
26. Hahn, E. L. (1949) *Phys. Rev.* **76**, 461.
27. Henke, B. L. (1964) *Advan. X-ray Anal.* **7**, 460.
28. Howling, H. D. (1965) *Rev. Sci. Inst.* **36**, 660.
29. Hughes R. H. (1947) *Phys. Rev.* **71**, 562.
30. Inghram, M. G., Hayden, R. J. H. and Hess, D. C. (1947) *Phys. Rev.* **72**, 349.
31. Jopson, R. C., Mark, H., Swift, C. D. and Williamson, M. A. (1964) *Phys. Rev.* **136**, A69.
32. Kemp, J. W. (1951) *Opt. Soc. America* (convention) Chicago 1951.
33. Kessel, W. and Jecht, U. (1964) *Z. Angew. Phys.* **17**, 283.
34. Klein, M. P. and Barton, G. W. (1963) *Rev. Sci. Inst.* **34**, 754.
35. Kogelnik, H. and Porto, S. (1963) *J. Opt. Soc. America* **53**, 1446.
36. Koningstein, J. and Smith, R. (1964) *J. Opt. Soc. America* **54**, 1061.
37. Leaf, R. H. and Jones, J. T. (1965) *J. Internationalisde Siderurgie.*
38. Leane, J. B., Richards, R. E. and Schaefer, T. P. (1959) *J. Sci. Inst.* **36**, 230.
39. Le Mieux, A. F. and Beeman, W. W. (1946) *Rev. Sci. Inst.* **17**, 130.
40. Nier, A. O. (1947) *Rev. Sci. Inst.* **18**, 398.

41. Norath, A., O'Sullivan, W. J., Robinson, W. A. and Simmons, W. W. (1964) *Rev. Sci. Inst.* **35**, 476.
42. Pound, E. M. (1950) *Rev. Sci. Inst.* **21**, 219.
43. Pound, E. M. (1951) *Phys. Rev.* **82**, 343.
44. Pound, R. V. (1946) *Rev. Sci. Inst.* **17**, 490.
45. Purcell, E. M., Torrey, H. C. and Pound, R. V. (1946) *Phys. Rev.* **69**, 37.
46. Radyne Ltd. Wokingham, Berkshire, England.
47. Reimann, R. (1963) *Compt. Rend.* **257**, 3862.
48. Ridenour, L. N. and Lampson, C. W. (1937) *Rev. Sci. Inst.* **8**, 162.
49. Ross, I. C. and Stafford, R. W. H. (1964) *Acta Imeko* 43.
50. Salmon, M. L. (1964) *Advan. X-ray Anal.* **7**, 604.
51. Shirley, D. A. (1965) *Nucleonics* **23**, 62.
52. Smith, F. and Rippon, K. (1962) *J. Brit. I.R.E.* **24**, 127.
53. Stallwood, B. J. (1954) *J. Opt. Soc. America* **44**, 171.
54. Strum, P. D. (1953) *Proc. Inst. Radio Engrs.* **41**, 875.
55. Torrey, H. C. (1949) *Phys. Rev.* **75**, 1326.
56. Weber, K. and Marchal, J. (1964) *J. Sci. Inst.* **41**, 15.
57. Weekley, R., and Norris, J. (1964) *Appl. Spectry.* **18**, 21.
58. White, J. W. (1967) *Rev. Sci. Inst.* **28**, 2, 187.
59. Williamson, G. K. and Smallman, R. E. (1954) *J. Sci. Inst* **31**, 68.
60. Yee, K. W. and Deslattes, R. D. (1967) *Rev. Sci. Inst.* **38**, 5, 637.

# 5 Electronics and Nuclear Measurements

## RADIATION DETECTORS

**The geiger tube**

Probably the most familiar radiation detector is the geiger tube and we shall consider this type first. The basic construction of an end window geiger tube and the outline of a geiger counter is shown in figure 5.1(a). An energetic charged particle from the radiation source, causes ionisation of the gas in the tube, electrons being attracted towards the central wire anode and slower moving positive ions to the tube wall. The electrons, in moving towards the anode collide with atoms of the gas filling and further ionisation, known as gas multiplication, is produced. When the applied voltage is such that the tube gives an output pulse which is proportional to the number of ions produced by the original particle, we are working in the proportional region. If the voltage is increased, so does gas amplification, and we operate in a region where the output pulse size is almost independent of the size of the original event. This is called the geiger region and if the voltage were increased still more, a continuous discharge would result which would damage the tube. This is prevented in geiger operation by the slow moving positive ions forming a sheath round the wire anode, which reduces the electric field near the wire and terminates discharge. It should be noted however, that until the positive sheath has reached a certain radius, the reduced electric field is insufficient to operate the geiger tube and it is insensitive for a period of about 100 $\mu$sec known as the dead time.

As the positive ions hits the tube wall they are capable of ejecting secondary electrons which could cause another pulse, and thus repeated pulses, so that it is essential to quench this second pulse. This may be done by (1) using a quenching gas (halogen or organic vapour) in which the energy of the positive ions and secondary electrons is dissipated without further ionisation being produced, or (2) by using an electronic circuit which will reduce the voltage applied to the tube for a time longer than the time required for the positive sheath to reach the tube wall.

**Gas flow counting**

Another method of geiger or proportional counting is that of gas flow counting. The principle of a gas flow counter is shown in

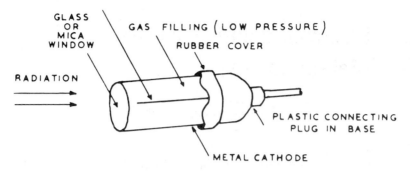

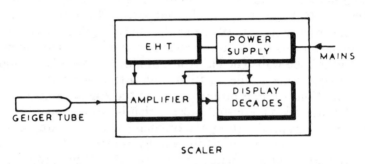

*Fig. 5.1.* (a) The Geiger counter.

figure 5.1(b), the detecting chamber having orifices at the top, through which a steady stream of gas flows at a slow rate into the chamber. This gas removes air introduced with the sample and the products of discharge, and may be Q gas for geiger operation or P.R. gas (90% argon, 10% methane) for proportional counting. A very thin window, which allows soft beta particles to pass with little absorption, fits onto the bottom of the counting chamber and prevents chamber contamination, electrostatic charge effects and vapour transmission from moist samples. This window may be expanded terylene (as used in the Beckman Low Beta counter) having a thin gold coating on both sides, the inner coating assisting in giving a uniform flat electric field, the outer preventing static effects.

Gas flow geiger type detector heads (using helium isobutane) are used on radiochromatography equipment. The detectors monitor each side of the radioactive chromatogram and are mounted on a scanner coupled with an electric typewriter. If the desired count is not reached in the set time the typewriter and scanner move on to the next position. If the count is reached in the set time then the typewriter prints out the count corresponding to the activity of the spot on the paper. Strip scanning systems are also available.

## Ionisation chambers

Another radiation detector is the ionisation chamber which consists basically of a gas filled chamber with two electrodes. A charged particle entering the chamber gives rise to an electron and a positive ion which are accelerated in opposite directions due to the electric field. If we consider the electrodes as capacitor plates, then the induced charge produces a change in the voltage across them which is measured by an electrometer valve. The ionisation chamber has been applied to gas chromatography, the principle being shown in figure 5.1(c). The radiation source is used to ionise the carrier gas into positive ions and electrons and if this gas is nitrogen, then nitrogen molecules and slow electrons are formed, the electrons migrating

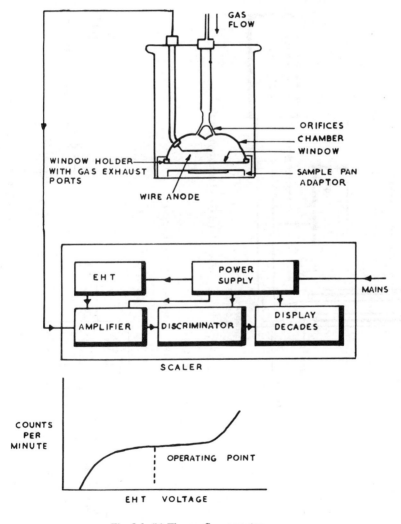

*Fig. 5.1.* (b) The gas flow counter.

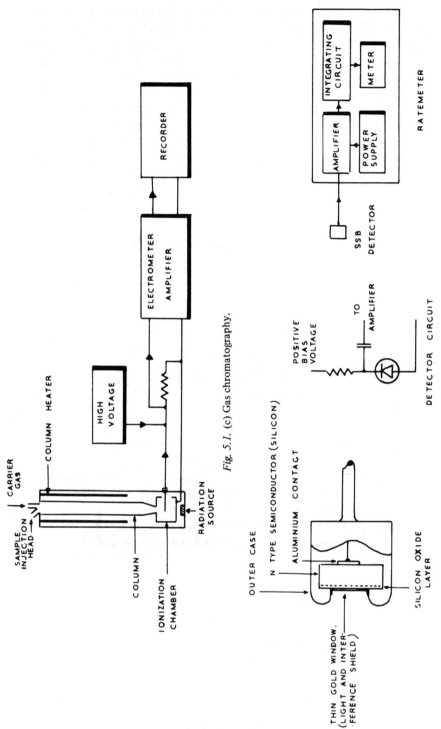

Fig. 5.1. (c) Gas chromatography.

Fig. 5.1. (d) The solid state radiation detector used as a ratemeter.

to the anode to produce a steady signal to the electrometer. If the
sample introduced contains electron absorbing molecules, then the
current is reduced, the loss of current being a measure of the electron
affinity of the compound. This technique is known as electron capture. Typical sources used in electron capture detectors are $H_3$ and
$Ni_{63}$. The former is usually temperature limited to 210°C and has
a half life of 12.5 years. Detectors using $Ni_{63}$ may have several
times the sensitivity of an $H_3$ detector and can be used up to at least
300°C. A comparison between the two sources is given below.

| Source | Source Strength | Beta Energy | Half Life | Minimum Sample Detectable Limit |
|---|---|---|---|---|
| $H_3$ | 130 mC | 0.018 MeV | 12.5 years | 1.8p grams. |
| $Ni_{63}$ | 10 mC | 0.06 MeV | 85 years | 0.6p grams. |

Another method, using a cross-section detector, utilises helium or
hydrogen as the carrier gas, any other vapour being admitted to the
detector producing an increase in current.


**Scintillation counting**

The most efficient method of detecting radioactivity is that used
in the scintillation counter. The detecting element is a photomultiplier tube in close proximity to either a solid or liquid scintillator,
which converts energy from the charged particles into light. Typical
solid scintillators are crystalline anthracine and thallium activated
sodium iodide, the anthracine having the highest light output of the
crystal organic scintillators. Portable contamination monitors may
use probes comprising a photomultiplier combined with a phosphor
element. Typical phosphors are zinc sulphide on perspex for detecting alpha particles, anthracine for beta particles, and thallium activated sodium iodide for detecting gamma radiation.

In liquid scintillation counting the sample is enclosed in a glass
vial and consists of the radio active material, a solvent and scintillation solutes. The energy of the particle is transferred to the solvent,
the excitation energy of which is then transferred to the molecules
of the primary solute. The excited molecules of the solute return
to the ground state and in so doing emit quanta of visible or near
infra-red light which is detected by the photomultiplier tube (27).
Secondary solutes may also be used and absorb the light emitted by
the primary solute and then re-emit it at a longer wavelength, so that
the light finally detected matches the wavelength for maximum
response from the photomultiplier. Table 5.1 shows the wavelength
of maximum emission of some primary and secondary solutes and
the wavelength for maximum quantum efficiency of various
photomultiplier photo cathodes (34) (40).

| SOLUTE | | PHOTOMULTIPLIER | | | | |
|---|---|---|---|---|---|---|
| Solute | Approximate wavelength of maximum emission | Photocathode | Example | S response | Wavelength for peak quantum efficiency | Typical peak quantum efficiency |
| *Primary* P.P.O. | 3800 Å | $C_s$–K–$S_b$ (Bi-alkali) | RCA 8575 | No 'S' designation peak response at 3850 Å | 3350 Å | 28–30% |
| p-terphenyl | 3500 Å | $CS_3$–$S_b$ | EMI 9514S | EMI S | 4000 Å | 16% |
| P.B.D. | 3600 Å | $CS_3$–$S_b$–O | EMI 6097 | S11 | 4000 Å | 18% |
| B.B.O.T. | 4350 Å | $CS_3$–$S_b$–O | EMI 6255 | S13 (quartz window) | 4000 Å | 18% |
| *Secondary* P.O.P.O.P. | 4200 Å | $S_b$–Na–K–$C_3$ | | S20 | 3600 to 4200 Å | 20% |
| Dimethyl P.O.P.O.P. | 4300 Å | $CS_3$–$S_b$–O | | EMI SUPER S11 | 4000 Å | 22% |
| α-N.P.O. | 4000 to 4200 Å | | | | | |

Notes: (1) The performance of a solute in a particular scintillation counter depends not only on the spectral match with the photomultiplier but on its concentration in solution, the scintillator formula, and the efficiency of light guides and reflecting surfaces used between the sample vial and photomultiplier window.
(2) Generally, photomultipliers with venetian blind dynode structures and $CS_3$–$S_b$–O photocathodes are used on cooled or controlled temperature scintillation counters. The latest ambient temperature systems use focused dynode bi-alkali photomultipliers.

TABLE 5.1.
*MATCHING OF PHOTOMULTIPLIER WITH EMISSION FROM LIQUID SCINTILLATION SOLUTE*

## Solid state detectors

The most recent type of detector is the solid state radiation detector shown in figure 5.1(d). This diagram shows the silicon surface barrier type and when a bias voltage is applied to the diode, formed by the semiconductor PN junction (silicon and silicon oxide), electrons in the N material are drawn towards the positive terminal while positive holes are attracted to the negative terminal. This results in a charge free region behind the junction, and radiation particles entering this region cause partial ionisation, electrons and positive holes being formed. The charge produced is directly proportional to the energy so that the detector is proportional and there is no dead time as with the geiger tube. Operation is very fast and the detector is suitable for low energy beta and alpha particles.

Table 5.2, compiled from published data summarises different types of radiation detectors. It should be noted however that efficiencies and background shown are approximate only and the information on liquid scintillation counters refers to ideal unquenched samples. The background count is given in counts per minute (c.p.m.) (see pp. 182-183).

## ELECTRONIC CIRCUITS

### The simple geiger counter

A simple unit of this type is the Panax 102ST scaler, a circuit of which is shown in figure 5.2. The output pulses from the halogen quenched geiger tube (MX168) are amplified by transistors Q1 Q2 and then coupled to the units dekatron driver by grounded emitter stage Q3, which delivers a 12 volt positive pulse to transistor Q4. This transistor and transformer T1 form a blocking oscillator, which delivers narrow width pulses to the guide electrodes of the dekatron tube. The positive pulse on point A causes Q4 to be conducting for about 60 $\mu$sec; the transistor then switching off and 120 volts pulses from the transformer secondary are fed to the dekatron guides. Transistor Q5 is normally conducting, but when the dekatron glow discharge moves to cathode 9 a positive voltage is fed to Q5 base turning it off. On the next dekatron step, the glow transfers to cathode 10 and Q5 is turned on again and the positive signal at its collector is used to drive the 10's dekatron unit. The output from this unit turns off transistor Q8 and turns on Q9 for a short time. This energises the coil of the mechanical register which indicates hundreds of counts, the coil remaining energised until capacitor C2 is discharged. The E.H.T. for the geiger tube is obtained from the blocking oscillator formed by Q10 and transformer T2. The output pulses from the secondary winding of T2 are fed to a voltage doubling rectifier circuit, its d.c. output being applied to the anode of the geiger tube. The output of the oscillator is controlled by transistor Q11 which controls the supply to Q10.

## GEIGER COUNTERS

| Type | Background (c.p.m.) | Particles Detected | Example | Performance (General) |
|---|---|---|---|---|
| Geiger tube in castle | 20 (shielded 2 in. lead) | Beta | MX 168 | |
| Geiger tube in castle | 3 (shielded 2 in. lead) | Alpha, beta | MX 113 | Approximately 6% efficiency for $C_{14}$ |
| Geiger probe | 45 (unshielded) | Gamma | MX 115 | |
| Geiger probe | 90 (unshielded) | Low energy gamma | MX 141 | |
| Shielded geiger tube | 50 (shielded 2 in. lead) | Soft X-rays | MX 118 | |
| Anti-coincidence | 2 (shielded) | Beta | MX 152 tube MX 157 guard counter and anti-coincidence circuit | |

## GAS FLOW COUNTERS

| Type | Background (c.p.m.) Shielded | Particles Detected | Example | Performance (General) |
|---|---|---|---|---|
| Planchet gas flow counter | 16 (beta proportional) 0.01 (alpha proportional) | Alpha, beta | Nuclear Chicago D47 (3 in. window diameter) used with suitable scaler | Approximately 27% efficiency for $C_{14}$ (35% with ultra thin window) Proportional gas requires high sensitivity preamplifier and detector operates at 2000-3000 H.T. Geiger gas gives large amplitude output pulses from detector which operates at 1000-1500 V H.T. |
| Low background planchet gas flow counter | 1-2 | Alpha, beta | Nuclear Chicago spectro shield (2 in. window diameter) system with anti-coincidence circuit | |
| Low background planchet gas flow counter. | 6 | Alpha, beta | Beckman low beta (5 in. diameter window) system with anti-coincidence circuit | |

| NaI (TL) Crystal | 240 | Gamma | Nuclear Enterprises 5502 system | Single photomultiplier liquid scintillation systems now superseded by two photo-multiplier instruments with fast coincidence and pulse summation circuits which enable a much higher figure of merit to be obtained (i.e. efficiency[2]/background). Typical figures of merit for modern systems are 100/150 for $H_3$ and 300/400 for $C_{14}$. In dual isotope counting a Klein-Eisler number of the order of 150-170 may be obtained, and less than 10-12% of the $C_{14}$ pulse spectrum falls in the $H_3$ channel when it is set for approx. 50% efficiency. Some instruments (e.g. Nuclear Enterprises 8310) also incorporate a gamma crystal. It should be noted that efficiencies and backgrounds quoted are approximate only and refer to ideal *unquenched* standard samples. |
|---|---|---|---|---|
| Liquid Scintillation single photomultiplier | 60 for 30% $H_3$ efficiency 300 for 40% $H_3$ efficiency | Beta | Nuclear Enterprises 8306 system | |
| Liquid Scintillation. Two photomultipliers and coincidence circuit. Temperature controlled system | 49 for 45% $H_3$ efficiency 33 for 79% $C_{14}$ efficiency | Beta | Nuclear Chicago 6725 system | |
| Liquid Scintillation. Two photomultipliers, coincidence and pulse summation circuits. Temperature controlled system | 20 for 55% $H_3$ efficiency | Beta | Nuclear Chicago 6860. Packard 3000, 4000 series. Nuclear Enterprises 8310. Intertechnique. Philips | |
| Liquid Scintillation. Two photomultipliers, coincidence and pulse summation circuits. Room temperature system | 40 for 37% $H_3$ efficiency 32 for 78% $C_{14}$ efficiency (differential counting) 30 for 57% $H_3$ efficiency | Beta | Nuclear Chicago 6850 Beckman L.S. 133, 150 | |

TABLE 5.2.
*CHARACTERISTICS OF RADIATION DETECTION SYSTEMS*

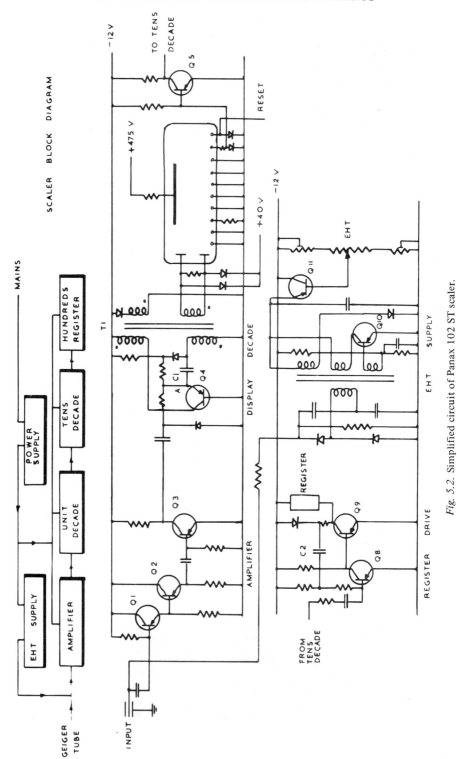

*Fig. 5.2.* Simplified circuit of Panax 102 ST scaler.

# ELECTRONIC CIRCUITS

Of course not all scalers use mechanical registers, many types having perhaps six transistor driven dekatron stages which are suitable for count rates up to 50,000 pulses per second. The geiger tube itself may also be enclosed in a lead castle, as is usual with four or six decade scalers, because this reduces the background counts obtained to the order of 20 cpm.

**The ratemeter**

The type of circuit described above will indicate a number of counts, but it is possible by using an integrating circuit to measure the count rate, an example of this type of instrument being the Panax RM202. Before describing its circuit operation we should first consider the basic integrator circuit consisting of a resistance and capacitance in parallel.

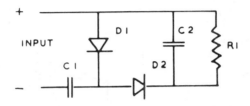

*Fig. 5.3.* (a) Ratemeter integrating circuit.

The random pulses from the detector are first shaped, as we are concerned with the count rate, and it is essential that each pulse should deliver the same quantity of charge to the input capacitor C1 of the circuit shown in figure 5.3(a). The shaped pulses applied to the input cause diode D1 to conduct and charge capacitor C1, which can then discharge through diode D2 and charge capacitor C2. With a steady rate of arrival of pulses, equilibrium is established when the rate of incoming charge balances the outgoing leakage from C2 through resistance R1. We thus develop an average voltage, proportional to the average counting rate, which may then be indicated on a meter. It should be noted that the time required to reach equilibrium depends on the time constant R1 C2 of the integrating circuit and also the count rate.

In the Panax RM202 rate meter circuit, shown as figure 5.3(b), negative pulses from the portable geiger probe are fed through an emitter follower Q1 to the blocking oscillator formed by transistor Q2 and transformer T1. The blocking oscillator produces an output pulse which turns transistor Q3 on, and a charge, determined by the range capacitor C1 in use, is fed to the rectifiers MR1, 2 and 3. The mean current carried by MR3 is proportional to the count rate and is smoothed by capacitor CS and fed to the amplifier Q4 Q5 Q6. The integrating time is determined by capacitor C2, which is in parallel with R1, this resistor carrying almost the whole of the input

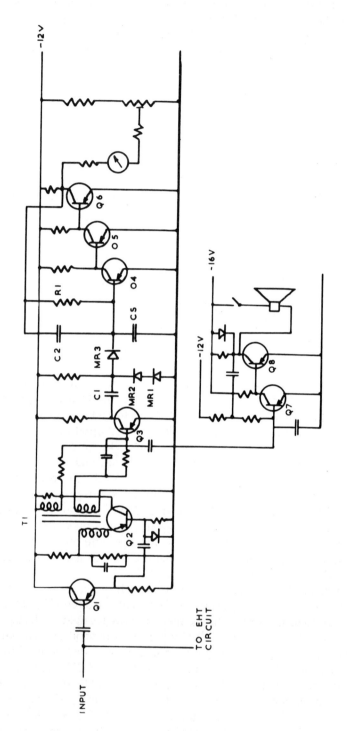

*Fig. 5.3.* (b) Simplified circuit of Panax RM202 ratemeter, note capacitors C1 and C2 are in fact capacitors selected by switches.

# ELECTRONIC CIRCUITS

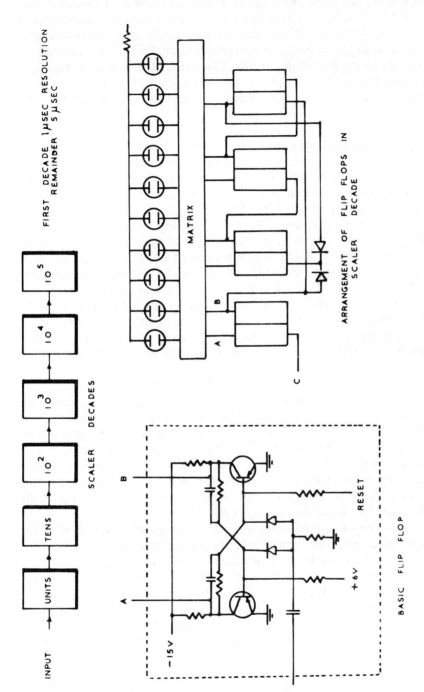

*Fig. 5.4.* Fast scaling circuit.

current (due to feedback from the collector of Q6 to the base of Q4). Audible indication of the count rate is given by a loudspeaker, the drive for this being a monostable circuit consisting of transistors Q7 and Q8. Transistor Q7 is normally on and Q8 off, so that a positive pulse from Q2 collector momentarily turns off Q7, its collector voltage rising so turning on Q8. This causes the supply voltage to be applied to the loudspeaker and results in a click being heard for each pulse. When the input pulse is absent the circuit reverts to its normal state of Q7 on and Q8 off.

### Scaling and display circuits

The simple geiger counter described above will only operate at a slow count rate (1,000 cps) due to the mechanical register, and for high speed scaling it is necessary to use a fast electronic circuit prior to the slow decades. A typical fast decade, with a resolution time of one microsecond, may utilise fast switching transistor flip flop or bistable multivibrator units. An outline of such a scaling decade is shown in figure 5.4, the flip flop circuit having two stable states (a) transistor Q1 fully on and Q2 cut off or (b) transistor Q2 fully on and Q1 cut off. If circuit component values are such that QI is normally fully on, then a positive input pulse applied to its base will tend to cause cut off, Q1 collector voltage going negative and in turn applying a negative voltage to Q2 base. Transistor Q2 now conducts, the action being cumulative until the second stable state is reached.

For one complete cycle of the flip flop circuit we have two input pulses, so that by connecting together four such binary circuits we obtain one output pulse for every sixteen input pulses. We wish however to count to a scale of ten, so it is necessary to apply feedback as shown between the flip flops in order to convert from a count of sixteen to a count of ten. If we represent the on state of the flip flop or binary right hand transistor by a 1 and the off state by a 0, we can construct a truth table which shows the state of the binaries for decimal counting (table 5.3).

If we connect nine neon lamps through a matrix circuit of diodes and resistors to the binary outputs, we can arrange that only one lamp is on at a time corresponding to the decimal input to the scaling unit. Figure 5.4 shows how several such units are connected together, the output of one decade feeding into the input of the next. Four outputs are taken from the decade multivibrator units and connections are made through a circuit matrix of resistors and diodes to the count display. The neon lamps are mounted behind transparent number indicators. Cold cathode indicator tubes may also be used to display the number of counts.

The "Nixie" digitron, numicator and nodistron tubes are gas filled cold cathode devices with an anode and ten in line cathodes shaped to the form of the numerals 0 to 9. When voltage is applied between the anode and appropriate cathode, a glow discharge takes place and the number is illuminated. The cathodes are driven by transistors which obtain their input from the matrix circuit of the scaler decade. This type of tube has an extremely long operating life, a figure of 70,000 hours under dynamic operating conditions being quoted (16) (17). In older equipment, glow discharge dekatron tubes were used for both counting and display, instead of multivibrator circuits with a separate visual display.

| Decimal input pulses | State of multivibrator units | | | |
|---|---|---|---|---|
| 0 | 0 | 0 | 0 | 0 |
| 1 | 0 | 0 | 0 | 1 |
| 2 | 0 | 0 | 1 | 0 |
| 3 | 0 | 0 | 1 | 1 |
| 4 | 0 | 1 | 0 | 0 |
| 5 | 0 | 1 | 0 | 1 |
| 6 | 0 | 1 | 1 | 0 |
| 7 | 0 | 1 | 1 | 1 |
| 8 | 1 | 0 | 0 | 0 |
| 9 | 1 | 0 | 0 | 1 |
| Multivibrator binary unit | 4 | 3 | 2 | 1 |

TABLE 5.3.
*SCALING DECADE TRUTH TABLE*

A scaling unit used on a modern multichannel beta or gamma counter may have six nixie type display tubes which can be switched to read the output of any scaler channel or the timer. The count capacity per channel would be 1,000,000 and a typical maximum count rate 20 megacycles. In addition to visual indication of the counting time and the counts obtained in each channel, there is usually provision for print out of the accumulated data.

The required number of counts and the counting time may be preset. The time pulses can be obtained from an oscillator circuit, the output of which is divided down by multivibrator units to 100 pulses per minute. These pulses are fed into a timer decade and transistor switching circuits are used so that when the preset count or time is reached, counting is stopped and, on automatic systems, the sample changed. Automatic multichannel instruments incorporate an electronic calculator with a printer to list the accumulated and calculated data. These systems also have outputs for punched tape or cards.

**Data print out**

Since the decade multivibrator circuits have outputs which are in binary coded form, it is necessary to utilise the conversion circuit

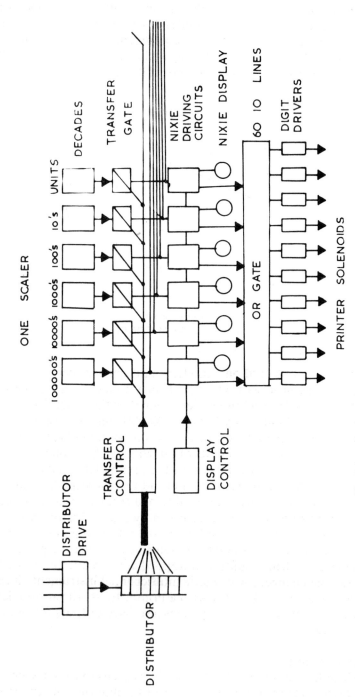

*Fig. 5.5.* Simplified diagram of print out system.

for decimal operation of the printer solenoids. Thus on a system using nixie tubes, outputs may be taken from the nixie driver transistors and applied to digit driver power transistors which in turn operate the printer solenoid. In another method, the emitters of all the 'odd' driver transistors are connected together, as are those of the 'even' transistors. These odd and even lines are then controlled by the first binary of the decade circuit, and final selection is by resistor-diode logic circuits at the base of each transistor.

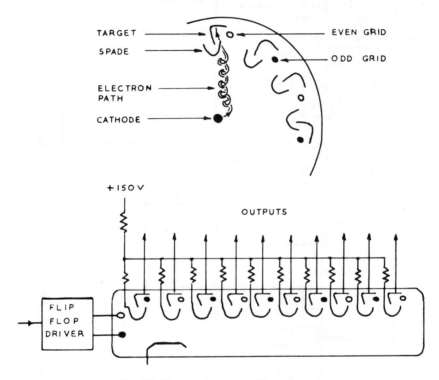

Fig. 5.6. Magnetic beam switching tube.

In a multichannel automatic counting system the information contained in the scalers, the counting time and the sample number must be fed in the correct sequence to the printer, and this may be done by using a 'distributor circuit'. Many different types of circuit are used to perform this function, ranging from uni-selector stepping switches with an internal dekatron type readout tube in older systems to a four layer Shockley diode distributor with logic circuits in a modern system. These circuits are too complex to be described in detail and vary from one instrument to another but figure 5.5 illustrates in a simplified form the basic ideas involved in one particular method.

On completion of counting, providing the printer and distributor are in the required condition, the distributor drive circuits allow

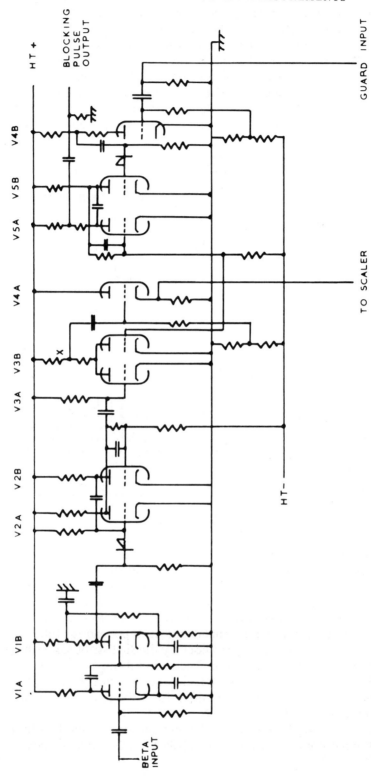

Fig. 5.7. Simplified circuit of Panax AU460 anticoincidence unit.

pulses from a 'clock' oscillator to step the distributor through its positions. The clock oscillator pulses are used to synchronise the various operations carried out. The distributor is connected to a control circuit which opens transfer gates connecting the four outputs of each binary coded decade to the nixie driver circuits, which convert the binary code to decimal. Each nixie circuit has ten outputs which are connected to a display tube, and also through a 60 to 10 line OR gate to the digit drivers which operate the printer solenoids. On the diagram, only one scaler is shown and with the distributor in the position to open the transfer gates for say scaler A, the states of the scaler A decades set up decimal numbers at the nixie driver output. The display control circuit allows the nixie drivers to deliver this information to the printer solenoids, and on completion of this operation the input circuits for the display control reset, and the distributor steps to the next position when the transfer gates for scaler B are opened.

**The magnetic beam switching tube**

Another device used for fast scaling is the magnetic beam switching tube shown in figure 5.6. The beam switching tube or trochotron, consists of a glass envelope containing a central wire cathode surrounded by ten target and spade electrodes, the whole assembly being enclosed by a cylindrical magnet. The electron beam from the cathode operates in crossed electric and magnetic fields, so that the electrons travel with a cycloid motion towards the positive targets. The individual electron velocity is thus very high but the beam rotational velocity low, so that by means of the crossed fields the beam can be directed towards a target, the positive spade electrodes helping to lock the beam on to the selected position. The beam is switched from one position to the next by lowering the potential applied to grid electrodes which are adjacent to the spades and targets, the switching time being of the order of 0.2 microseconds. The grids are designated odd and even, all the odd grids and all the even grids being grouped together and each group connected to the output of a multivibrator which is triggered by a decade drive circuit. The beam switching tube may precede flip flop or dekatron decades and can be used to operate nixie type indicators directly, thus eliminating many transistors and diodes.

The operation of valve and solid state counting and scaling circuits has been described in detail by Dance. (15)

## REDUCTION OF BACKGROUND COUNT

**Anti-coincidence circuits**

The background count, measured on a scaler operating from a geiger tube in a lead castle, is of the order of twenty counts per minute. This figure can be reduced to approximately two counts per minute by using a guard detector and an anticoincidence circuit. The guard detector tube is highly sensitive to cosmic radiation and if we obtain an output pulse from both the geiger and guard tubes, then the anticoincidence circuit prevents any count being registered on the scaler. The scaler thus only counts pulses from the geiger tube which do not coincide with pulses from the guard tube.

An example of this type of circuit is the Panax AU460 anticoincidence unit which utilises a Mullard MX152 beta geiger tube and a Mullard MX157 guard tube. The circuit is shown as figure 5.7. A 2 millisecond blocking pulse is applied to the beta channel whenever the guard tube discharges, and the beta channel is therefore inoperative for a fraction of the counting time dependent on the number of guard counts detected. The beta channel itself has a recovery time of 250 microseconds, so when the count rate is high, the observed beta counts should be corrected for losses due to guard blocking pulses and for losses due to beta channel dead time.

Negative pulses from the beta counter are passed through the amplifier V1A to the fixed bias discriminator V1B, which gives an input sensitivity of 0.5 volts. The output pulses from V1B anode trigger the flip flop V2 which introduces a dead time of 250 microseconds. Negative pulses from the trailing edge of V2 output are fed to V3A so that, providing V3B is not conducting, positive pulses occuring 250 microseconds after the beta input pulse are produced at V3 anode, and fed via the output cathode follower V4A to the beta scaler. Positive pulses exceeding 20 volts from the guard tube (due to its low resistance being connected between cathode and earth) are inverted by V4B and trigger the two millisecond flip flop V5. This causes V3B to be made conducting for two milliseconds and during this time any negative pulses on V3A grid will not give rise to any beta output pulses. This is because both halves of V3 are now conducting, and the voltage at point X is such that the cathode follower V4A cannot pass pulses to the output. A second scaler may also be used to register the blocking pulses and the output for this is taken from V5A anode.

Anticoincidence circuits are also used on low background gas flow planchet counters, two examples of which are the Beckman low beta and Nuclear Chigago Spectro-shield automatic systems. In the Nuclear Chigago equipment, a three inch diameter photomultiplier optically coupled to a plastic scintillator is used as the cosmic ray guard detector, and with a transistorised anticoincidence circuit can reduce the background to approximately one count per minute. The material used in low background systems must be specially selected, otherwise alpha and beta particles would be emitted from the detector materials and increase the background. It is also usual to surround the detector with three or four inches of virgin deep mined lead for gamma ray shielding and to use petrochemical plastics in which the carbon-14 has decayed to carbon-12. Gold, silver and copper are also used and are very pure so that any radioactivity present is of short half life.

**Coincidence circuits**

The reduction of background is particularly important in liquid scintillation spectrometers counting low energy beta emitting isotopes. A single photomultiplier scintillation counter operated at normal ambient temperatures will give background pulses (due mainly to thermal emission from the cathode) of the same order as the output obtained when measuring tritium. In order to reduce the number of thermal electrons, the photomultiplier may be cooled as thermal emission is then reduced by approximately a factor of five for every 13°C, at pulse heights corresponding to single electron emission (41). With a modern S11 CsSbO cathode, this thermal dark current is about 300 electrons/cm$^2$/sec at 20°C which gives a rate of 5,000 electrons per second in a typical 2" tube. The E.M.I. S cathode has a much lower thermal emission (20 electrons/cm$^2$/sec at room temperature), although the quantum efficiency is somewhat lower than in an S11 tube. Quantum efficiency is defined as the ratio of photo electrons emitted to incident photons. The new high quantum efficiency E.M.I. super S11 cathode has a thermal emission of about 8,000 electrons per second in a 2" tube at 20°C and the typical electron count rate/sq cm can be reduced from 100 to 10 by cooling from 20°C to 0°C. Bialkali tubes have a low dark current, the cathode giving only perhaps 300 or 400 electrons per second in a 2" tube at 25°C. This type of photomultiplier is therefore very suitable for use in ambient temperature counting systems. (44).

Photomultiplier cooling is achieved by passing water or a refrigerant through cooling coils in the detector head or by enclosing the photomultiplier and sample changer assembly of an automatic system in a refrigerated cooling unit. The photomultiplier thermal noise pulses are random events and this contribution to the background count can be considerably reduced by using two photomultipliers and a coincidence circuit, which only allows a pulse to be counted if a scintillation is simultaneously detected by both photomultipliers. With the advent of very fast coincidence circuits having a resolving time of 20 to 70 nanoseconds and high quantum efficiency low thermal emission bi-alkali photomultipliers, high counting efficiencies and low background can be obtained with scintillation counters operating at ambient temperatures. A simple transistorised coincidence circuit is shown as figure 5.8(a). Transistors Q1 and Q2 are normally cut off or non-conducting due to the positive bias from battery B1. However, when large negative pulses are applied to inputs A and B, both transistors conduct and current flows through the output resistance R3. If only one input pulse is applied, say to input A, then transistor Q1 has a large negative signal on its base but no current flows as the other transistor is still off. If the input signals at A and B are derived from photomultiplier tubes then an output from the circuit is only obtained if both photomultipliers simultaneously detect a scintillator light flash. Very fast coincidence circuits however, do not use ordinary transistors but may instead utilise tunnel diodes.

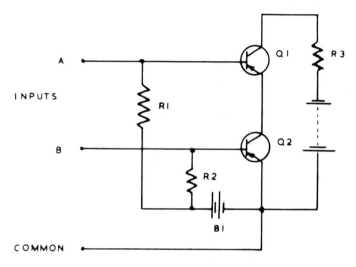

*Fig. 5.8.* (a) Basic transistor coincidence circuit.

A simplified circuit of the tunnel diode coincidence system used in the Nuclear Chigago 6860 (mark 1) and bench top 6850 (unilux) systems is shown as figure 5.8(b). Transistors Q1 and Q2 are normally conducting, the tunnel diode current flowing through them, the bias point for the tunnel diode being shown as point A. If we have an output from photomultiplier tube PM1, then a negative going pulse appears at the amplifier output and cuts off the diode D1 at the input to transformer T1. The transformer primary current is thus interupted and a positive spike is produced and applied to transistor Q1 base. Transistor Q1 is then cut off, but the tunnel diode bias point is such that unless we have outputs from both photomultipliers and both Q1 and Q2 are cut off, we cannot trigger or flip the tunnel diode to its 'high' voltage 'low' current state. Thus with coincident photomultiplier pulses, we cause a 0.5 volt tunnel diode output pulse to be applied to transistor Q3. This transistor in turn delivers a positive pulse to the amplifier formed by transistors Q4 and Q5, the output of Q5 being applied to a gating circuit in the main amplifier.

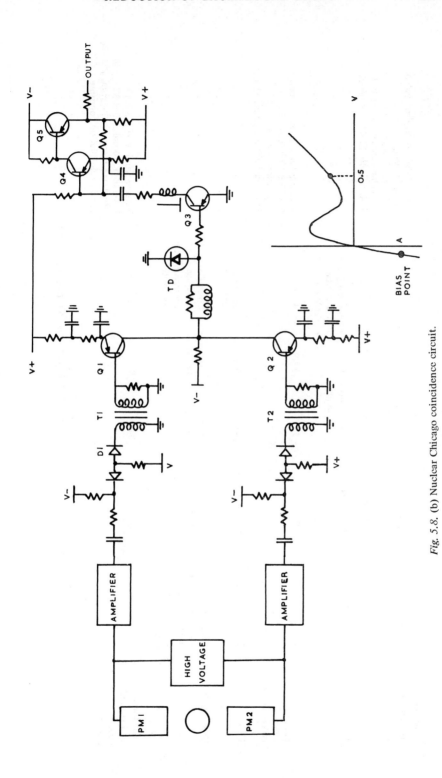

*Fig. 5.8.* (b) Nuclear Chicago coincidence circuit.

| | | | | | | |
|---|---|---|---|---|---|---|
| % of total background count | 41.3 | 8.4 | 27.3 | 23 | | Single photomultiplier system (Ekco N664). Wide window (5 v-∞). Total background count 143 c.p.m. [Ekco Tech 'Bull' S19 1961] |
| From | Photomultiplier | Sample carrier | Vial | Phosphor | | |
| Due mainly to | Thermal noise | Materials | Phosphorescence. K40 in glass | Cosmic radiation. Environmental radioactivity | | |
| % of total background count (wide window) | 3.2 | 18 | 17.6 | 12.9 | 48.3 | Two photomultiplier system (Nuclear Chicago 725). Wide window (0.5 v-∞). Total background count 89 c.p.m. |
| From | Photomultipliers | Photomultipliers | Vial | Solvent (toluene) | Scintillator (PPO POPOP) | |
| Due mainly to | Thermal noise | Thermal noise (counting chamber blocked so tubes cannot see each other) | "Light dark current" | K40 in glass | Cerenkov radiation | Cosmic radiation. Environmental radioactivity |
| % of total background count (narrow window) | 4.8 | 31 | 39.5 | 9.2 | 15.5 | Narrow window (0.5 v-5 v). Total background count 33 c.p.m. [Nuclear Chicago Publication 711580] Liquid Scintillation counting. |

TABLE 5.4.
LIQUID SCINTILLATION COUNTER BACKGROUND COMPOSITION

## Liquid scintillation counter background

The background count obtained in a liquid scintillation system depends on many factors, not just the thermal noise pulses from the photomultiplier. We have previously mentioned the need for minimising natural radioactive material in a counter and this also applies to the photomultiplier end window and sample vial. Cosmic radiation, phosphorescence in the sample vial glass caused by exposure to the ultra violet wavelengths in daylight or artificial light, potassium in the sample vial glass, chemiluminescence caused by chemicals added to the solution to increase sample solubility, electrical 'noise' in the counting system and internal photomultiplier effects are also contributors to the background count (45) (38). The internal photomultiplier effects which may give rise to spurious noise pulses are light flashes inside the tube due to electron avalanche, and positive ions produced by electron bombardment being attracted to the cathode (42) (10). Table 5.4 shows the background composition of a single photomultiplier counter and a two tube coincidence counter (19) (35). In a modern liquid scintillation counter set up for counting tritium, about 40% of the background count may be attributable to potassium in the sample vial, a further 40% to internal photomultiplier effects and approximately 20% to cosmic radiation.

In single photomultiplier counters, the background due to thermal noise may be reduced to a low level by refrigeration down to $-10°C$ or $-20°C$. It is then necessary to add absolute alcohol or some other anti-freeze to the sample in order to depress its freezing point. (20). Not all single photomultiplier counters however operate at this low temperature as acceptable results can be obtained by operation a few degrees above freezing point. Phosphorescence of sample vials may be troublesome in the single tube system, and it is necessary to dark adapt for perhaps 20 minutes before taking measurements. The vials should not be exposed to ultra violet rays or the residual phosphorescence will increase to a high level (4). This condition may remain for several hours unless alleviated by heating to about $70°C$ followed by cooling in darkness. Filters made from Chance OX7 glass may be used to overcome phosphorescence effects. (31).

The two photomultiplier coincidence method rejects pulses caused by phosphorescence and chemiluminescence, as there are single photon events and multiple photon events are necessary to obtain coincidence pulses from both photomultiplier tubes. The chance or accidental rate R for a two photomultiplier system is given by $2.R1.R2.T$, where R1 and R2 are the individual tube noise rates and T the coincidence resolving time (47) (37). There is a practical limit to the resolving time due to the time period of the photomultiplier output pulse. This is related to the rise time of the photomultiplier (about 2 nsecs for focused dynode and 7 nsecs for venetian blind structures) and the time of decay of fluorescence which is generally in excess of 3 nsecs (46) (32).

The total background count in a modern liquid scintillation counter, using pulse summation and a fast coincidence circuit, may be 20 cpm for an unquenched tritium efficiency of nearly 60%. Of this background, only one cpm or less originate from chance or accidental coincidences. Short resolving times are of advantage in ambient temperature systems where the effect on the thermal noise of increased temperature may be overcome by the short resolving time. It has been shown however that the single tube noise rate from a 6097B photomultiplier at $+5°C$ is of the same order as the rate from a 6097S tube at $+27°C$ (23). Thus the accidental coincidence rate of a refrigerated system using standard tubes is comparable to that obtained in an ambient temperature system using tubes with cathodes specially processed for low thermal emission. Bi-alkali photocathodes with their low dark current and high quantum efficiency are very suitable for ambient temperature scintillation counters.

## PULSE AMPLIFICATION

### Proportional amplifiers

In proportional counting equipment the radiation detector gives an output electrical pulse which is proportional to the energy of the incident radiation, and to obtain a measurement of this energy it is important that the amplifier characteristic is stable.

Where the capacitive effect of the cables between the detector and scaler would be detrimental, pre-amplifiers adjacent to the detector are necessary. This technique may be used in scintillation counters and in gas flow counters where the pre-amplifier allows the use of long cables between the detector and scaler. The pre-amplifier itself usually has one or two amplifying stages followed by a cathode follower, its output being coupled to the scaling unit.

### Pulse Spectrums

Before looking at some of the problems of amplifying the photo-multiplier output pulses, their nature must be considered. The particles emitted from a beta source vary continuously in energy and give rise to a spectrum of the general shape shown in figure 5.9(a). Gamma sources, unlike the beta emitters, do not give a continuous spectrum but a series of peaks. Since the beta spectrum of each particular isotope which we may wish to measure on a counter, will give pulses that range in size from nearly zero to maximum energy, it can be seen that the amplifier must be capable of handling a wide range of pulse heights. The height of the electrical pulse obtained is proportional to the energy of the beta particle that produced the pulse, and the proportionality factor is a function of the gain or amplification of the system. The overall system gain can be varied by altering the voltage applied to the photomultiplier, or by the use of an attenuator in the amplifier. The effect of this on the output pulse spectrum is shown in figure 5.9(b), and since we must assume that the activity of the radioactive sample has not changed in the measuring period, the area under the curve is constant. If it is arranged that we will only count pulses between two pulse height levels L1 and L2, then with the high voltage at a value HV1 we would measure practically no counts, as the majority of the pulses are below L1. If the high voltage is now changed to HV4, we have the condition (known as balance point counting) when the counts in the window L1 to L2 are at a maximum.

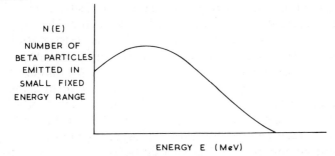

*Fig. 5.9.* (a) Beta spectrum.

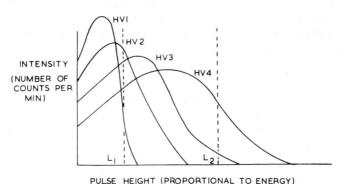

*Fig. 5.9.* (b) Pulse spectrum.

Most liquid scintillation systems are linear, the photomultiplier giving an output pulse directly proportional to the beta particle energy. As stated previously the range of pulse heights produced is large and may be 2,000 to 1. A single amplifier cannot cover this range of input pulses satisfactorily so that several linear amplifiers are used. An alternative is to use a logarithmic amplifier to compress the dynamic range although this type of amplifier is more difficult to keep stable. Another approach is to operate the photomultiplier in a pseudo logarithmic manner so that the range of pulses is approximately 100 to 1 for a 2,000 to 1 dynamic range of radiation energy.

**Amplifier characteristics**

The requirements for the scintillation counter amplifier are (1) good stability and freedom from drift (2) good linearity (3) fast response (4) fast recovery time (5) overload tolerance (6) low noise and (7) high gain. A large overload pulse applied to an amplifier will cause it to saturate, and a temporary paralysis may follow which is referred to as blocking. During this period, any further pulses are lost and if the recovery time is slow, the measurement of a pulse before circuit conditions have reverted to their original state, will be distorted. The main cause of blocking which occurs in a valve amplifier is the valve grid being driven positive, the resulting grid current charging the coupling capacitor. The capacitor will, of course, discharge through the grid leak resistor after the recovery of the input pulse, but if the RC time constant is large, the amplifier will have a long recovery time. A number of methods are used to overcome this problem, among them being direct coupling between stages, the use of diodes which conduct during large amplitude pulses so limiting the voltage swing and amplifier stages with a common cathode resistance (11) (21). In transistorised circuits which are driven beyond cutoff or saturation, the coupling capacitor may become charged through a low impedance and discharge through a high impedance, although this can be eliminated by suitable biassing (3) (18).

An undesired electrical signal, or noise as it is called, will effect the accuracy of measurements, especially of low level signals. Thermal noise signals can never be removed but may be minimised by the use of suitable operating conditions (49). Electrical noise may originate from the a.c. supply input, the counter itself, or radiation from adjacent electronic equipment. Resistance capacitance line filters are used to reduce the effect of supply line transients. Good earthing is of course essential.

A full discussion of the characteristics of nuclear pulse amplifiers is not possible in a book of this nature, but mention should be made of the change of shape of the pulses as they progress through the amplifier. If a pulse arrives before the preceeding pulse tail has decayed, a series of steps is produced and this would lead to a progressive rise of the amplifier input voltage and system paralysis, unless the signals are differentiated. If only one differentiating network is used, its position is very important, as placing the network at the amplifier output enables pulse pile up to occur, although any noise developed in the early stages is attenuated. By placing the network at the front of the amplifier, pile up cannot occur but any noise in the sensitive first stages is amplified.

Another disadvantage is the fact that the base line varies with the counting rate. These disadvantages can be overcome by using two differentiating networks so that double differentiation is obtained. This gives bipolar pulses and in a linear amplifier the positive and negative areas are equal and the base line stable (22) (12).

Typical performance figures for a scintillation counter linear amplifier are as follows and serve to illustrate the above mentioned points.

Gain: 10 to $10^3$

Recovery time: 5 microseconds after 200 times overload, 25 microseconds after 4,000 times overload

Linearity: 0.1% (although ± 1% is quoted for some beta spectrometers)

Stability: better than 0.1% between 0 and 55°C

Rise times on modern transistorised amplifiers vary from 10 nsecs to 160 nsecs.

For stable operation it is necessary to regulate to a high degree the high voltage applied to the photomultiplier and also the power supply to the transistor circuits. Typical regulation figures are 0.0005% change in transistor LT voltage and 0.0003% change in photomultiplier high voltage for a 1% change in line volts between 105 and 125 volts.

## PULSE HEIGHT ANALYSIS

In order to differentiate between particles of different energy, we have electronic circuits called discriminators which give an output when the input pulse height exceeds a certain preset value. The discriminator may be integral with the amplifier, a section of which has an adjustable bias, so that pulses less than a certain level are rejected while those that exceed this level are passed on to the display decades. Several types of electronic circuits are used as discriminators, two common examples being the Schmitt trigger and the tunnel diode discriminator.

### The Schmitt trigger discriminator

The basic Schmitt trigger circuit is shown in figure 5.10, and is one of the large family of multivibrator circuits. If in the initial state transistor Q1 is off, then a negative voltage is coupled to the base of Q2 which is turned fully on. This is due to the emitter-base junction being forward biassed. A negative going pulse applied at the input, sufficiently large to overcome the reverse bias on Q1, will turn this transistor on. The collector voltage of Q1 then decreases, the change being coupled to Q2 base. The emitter current of this transistor then decreases so that a regenerative circuit action occurs, resulting in Q1 being turned fully on and Q2 off. This state is maintained until the input pulse becomes positive going, which results in a reverse action occurring causing the circuit to revert to its initial state. The output pulse taken from the collector of Q2 is a square wave, its duration depending on the length of time the input pulse exceeds the circuit triggering voltage. It should be noted that there is a circuit hysteresis effect, giving rise to a slight difference between the increasing and decreasing threshold switching voltages. The voltage at which the circuits will trigger and turn Q1 on can be varied by applying bias (via resistance Rb) from the potentiometer L, which thus sets the value of pulse heights that are to be counted. In order to increase the sensitivity of this

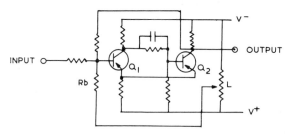

*Fig. 5.10.* Basic Schmitt trigger discriminator.

type of discriminator, the Schmitt circuit may be preceeded by a difference amplifier. Diodes may be used for preventing excess bias being applied to Q1 base and also to keep the transistor from becoming saturated.

### The tunnel diode discriminator

The tunnel diode may also be used as a discriminator. (figure 5.11). If the device is biassed to point A on the curve, a large current pulse at the input will cause the operating point to be flipped to point 1. This results in the tunnel diode circuit giving an output pulse of perhaps half a volt. By altering the bias to point B, a much smaller input pulse will cause the tunnel diode to flip to point 1. Thus, by setting the bias current with our level control we can use this device as a discriminator. The tunnel diode peak current is extremely stable with temperature and its switching time time very fast so that the discriminator has a low dead time and good stability (50) (24).

### Multi-channel analysis

Using discriminator circuits for pulse height analysis, by directing the amplifier output pulses into channels, we can build up spectrometer systems. This technique has been applied to scintillation counters using both crystal and liquid scintillator radiation

# PULSE HEIGHT ANALYSIS

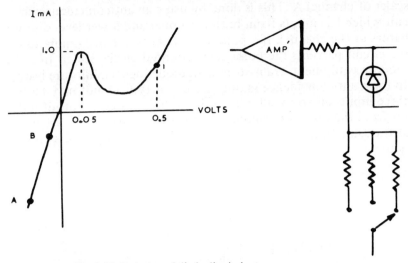

*Fig. 5.11.* Basic tunnel diode discriminator.

detection systems. A liquid scintillation system having 3 discriminators (figure 5.12), can be set up so that level L1 prevents noise pulses being counted, and levels L2 and L3 allow tritium initiated pulses to fall mostly in channel A (L1 – L2) and carbon 14 initiated pulses to fall mostly in channel B (L2 – L3). There is of course some overlap of the pulse spectra in these two channels. The pulses obtained from the discriminator units may be further shaped and the rectangular constant amplitude pulses are fed to the scaling circuits for counting.

Pulses which trigger discriminators L1 and L2 but not L3 should be counted on the scaler of channel B (L2 – L3) but not on the

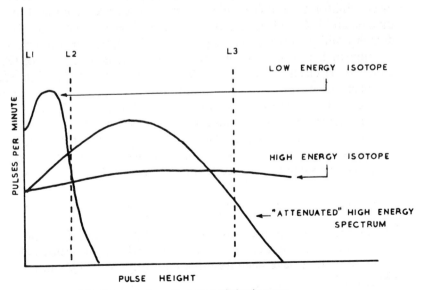

*Fig. 5.12.* Pulse spectrum for beta-emitting isotopes.

scaler of channel A. This is done by using an anticoincidence circuit which has inputs from both the upper and lower level discriminators of the channel. If a pulse triggers both the discriminators, the circuit prevents the pulse being counted on the scaler. In modern equipment, transistors with diode logic circuits are used to form anticoincidence gating circuits. A typical unit may have three inputs and only allows a pulse to be counted if output pulses from the lower discriminator and coincidence circuit are present at the same time, but without an input from the upper discriminator.

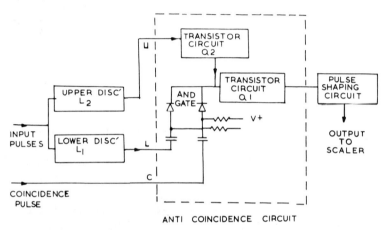

*Fig. 5.13.* Simplified anticoincidence circuit.

An anticoincidence circuit of this type, used in Nuclear Chigago equipment, is shown in figure 5.13. The diodes are normally forward biassed and form an AND gate. Negative pulses from the lower discriminator L1 and coincidence circuit C will reverse bias the diodes, and the AND gate output then enables transistor Q1 to deliver a pulse to the pulse shaping circuit. Should the upper discriminator U be triggered, its output pulse turns on transistor Q2 which short circuits the input from the AND gate to Q1. There is therefore no output from the circuit and the pulse is not counted.

The analyser section of a multi channel liquid scintillation spectrometer may have six independent discriminators, the voltage level control either being continuously variable or set in small voltage steps. The range covered may be from nearly zero to say 10 volts, the system linearity being better than 1%.

## LIQUID SCINTILLATION COUNTER SPECTROMETERS

The liquid scintillation counter is becoming increasingly important in the life sciences and the manner in which the various electronic units that have been described are utilised is shown in figure 5.14.

# LIQUID SCINTILLATION COUNTER SPECTROMETERS

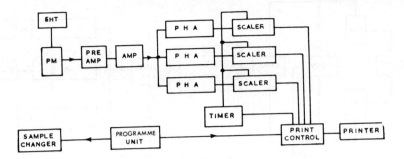

*Fig. 5.14.* (a) Simplified block diagram of Nuclear Enterprises 8305 system.

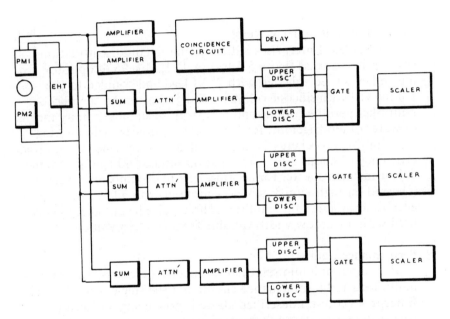

*Fig. 5.14.* (b) Analyser section of Nuclear Chicago Mark I Model 6860 System.

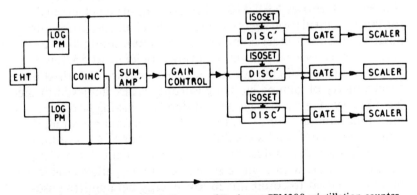

*Fig. 5.14.* (c) Simplified block diagram of Beckman CPM200 scintillation counter.

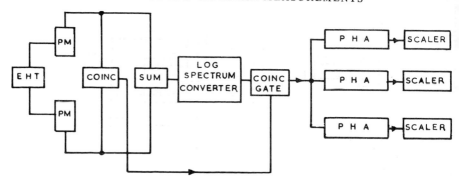

*Fig. 5.14.* (d) Simplified block diagram of Ansitron II scintillation counter.

## Single photomultiplier systems

The Nuclear Enterprises 8305 automatic liquid scintillation counter is shown as figure 5.14(a). A 1" E.M.I. 9524S photomultiplier is used shielded with 2" of lead and coupled to the sample vial with silicone oil. A cooling jacket round the photomultiplier permits cooling and temperature stabilisation by means of water or a refrigerant. Reflector coated quartz vials are used for counting low energy isotopes such as tritium when a counting efficiency of about 30% with a background of 60 cpm is obtainable using mains water for cooling. The photomultiplier output is passed through a unity gain pre-amplifier with a low output impedance to a variable gain amplifier. Three pulse height analysers are used with a window width variable from 0.1 to 5 volts.

## Performance figures

Most modern liquid scintillation spectrometers use two photomultipliers and a coincidence circuit and the single tube Nuclear Enterprises counter described above is now being replaced by a new system using these techniques.

As an example of a modern system we will describe the new Nuclear Enterprises NE 8310 automatic sample changer system for liquid scintillation spectrometry. Two bi-alkali photomultipliers are used with fast coincidence and pulse summation circuits to give a typical tritium counting efficiency of 55%, with a background of 20 cpm for unquenched samples. Scintillation spectrometry of gamma emitting isotopes is also achieved in the NE 8312 system by using a second detector having two photomultipliers and a 2" or 3" NaI crystal. The crystal has a radial sample access hole and a resolution of 10% for $^{137}$Cs is obtained.

The facility of two detectors in one instrument (NE 8312), each compatible with a single sample container, allows the instrument to be used to count mixed low energy beta emitting isotopes, mixed isotope gamma samples and mixed beta-gamma samples. The sample vials are loaded in racks of 10 in trays the instrument having a total capacity of 400 samples.

Modern high performance liquid scintillation spectrometers may employ racks and trays for sample handling (as in the above system), rotating turntables or conveyor belts and chains. The majority of systems have three counting channels, employ separate channels for external standard quench correction, and have facilities for data handling and print out of data. Spectrometers may also have built-in computation circuits or may be coupled to an external computer for quench correction, experimental and statistical calculations. The Nuclear Enterprises system is an example of the latter type and can be coupled to an Olivetti Programma 101 desk top computer. Punched tape outputs are also available.

The performance of a scintillation counter is indicated by a figure of merit which equals the counting efficiency$^2$/background. Typical values claimed for current systems are 150 for unquenched tritium and 350 for unquenched carbon 14. A new performance figure, which takes into account the degree of spectral separation in a scintillation counter has been described by Klein and Eisler (29). It should be noted however, that the efficiency of a particular solute in a scintillation counter depends on several factors, including the concentration in solution, the scintillator formula, degree of quenching and the optical "coupling" between the vial and photomultiplier (aluminium or titanium dioxide reflecting surfaces or plastic light guides).

Ambient temperature counting systems have the advantage that sample solubility is increased. Some solvents however (e.g. dioxane, which is water miscible) may decompose easily at ambient temperatures and for this type of sample, cooling below $+15°C$ is desirable in order to avoid high background. It should be noted that dioxane has a freezing point of $+12°C$. If a controlled sample chamber is used the temperature is usually variable over a range of $+10°F$ to $+50°F$ with a controlling accuracy of $\pm 1°F$ from the set point within this range.

**Pulse summation**

In a two tube sytem the light output from the sample is divided between the tubes. If we assume that four photons are produced by a disintegration, and the event is seen by both tubes equally, then two photons arrive at the cathode of each photomultiplier. This is a disadvantage compared to the single tube system where it can be arranged that nearly all the light produced falls on the photomultiplier cathode. This disadvantage is however overcome by the technique of pulse summation, in which the two outputs are added together in an electronic circuit prior to being passed to the analyser section of the counter. It must be remembered that we are adding coincidence pulses so that the signal height or amplitude is increased. Noise pulses however are random and this means that the overall signal to noise ratio is improved by pulse summation. There is a marked difference in the spectra obtained with and without pulse summation as in the former case the pulse height is larger and the spectrum is shifted toward the right. (48). The method also gives a slightly increased counting efficiency, as some of the lower energy pulses are raised above the level of the lower discriminator setting. Better separation of spectral peaks is also obtained and this is advantageous in dual labelled counting.

**Linear systems**

A block diagram of the two tube, three channel, linear, Nuclear Chigago Mark 1 refrigerated liquid scintillation spectrometer is shown as figure 5.14(b). This system has a coincidence resolving time of less than 70 nanoseconds and utilises amplifiers and discriminators having linearity better than 1%. There are six independent tunnel diode discriminators adjustable in 0.1 volt steps from 0.3 to 9.9 volts. Venetian blind type photomultipliers are used and also operate in a linear manner. Similar linear systems are available for instance from Nuclear Enterprises, Packard and Tracerlab Ltd. All of these instruments use up to 3 separate channels for sample measurement each having its own amplifier and upper and lower level discriminators.

**Logarithmic systems**

The Beckman ambient temperature system referred to earlier and shown in figure 5.14(c), uses focussed dynode photomultiplier tubes which have a fast electron transit time from cathode to anode. This enables a very fast coincidence circuit to be used with a resolving time of 20 nanoseconds. The photomultiplier tubes have bi-alkali photocathodes with a quantum efficiency of the order 30% and are operated so that their energy output is essentially logarithmic. Only one amplifier with a variable gain control is used.

In logarithmic systems there are two basic methods of acheiving a compressed dynamic pulse height range. The first uses a logarithmic amplifier after the pulses have been added together (Picker Ansitron and Intertechnique) figure 5.14(d) and the second uses photomultiplier tubes operated to give a pseudo lagarithmic characteristic (Beckman). The spectra obtained on a logarithmic system are different in shape to those encountered on a linear instrument and have steeper trailing edges. The discriminators which may be positioned on a trailing edge, therefore have to be very accurate, and rate meters may be used in conjunction with the discriminators in order to facilitate setting up the instrument. A useful feature of these systems is the utilisation of plug in 'isoset' or 'betaset' modules which contain factory adjusted components fixing the voltage levels of the discriminator units. This enables the operator to plug in the module for the isotope being counted.

**Live timing**

Some systems employ live timing, in which a coincident pulse initiates a constant delay period, during which any further pulses applied to the analyser are blocked. The clock time equals live time plus the dead time. In clock time systems corrections must be made at high count rates for the dead time. This is not necessary in systems employing live timing, which also overcomes the problem of satellite pulsing. That is, a small pulse occuring after a large pulse, due to slow photon release or anode feedback (light flash at anode) in the photomultiplier.

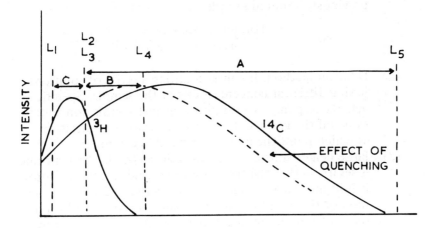

$$\text{Tritium ratio} = \frac{L_3 \text{ to } L_4}{L_1 \text{ to } L_2} = \frac{B}{C}$$

$$\text{Carbon ratio} = \frac{L_3 \text{ to } L_4}{L_3 \text{ to } L_5} = \frac{B}{A}$$

Dotted $^{14}$C curve shows effect of quenching

*Fig. 5.15.* Channels ratio.

**Quenching**

The presence of certain substances (e.g. oxygen) in the solution may interfere with the transfer of energy from the radioactive sample to the solvent, and this is called chemical quenching. Another form of quenching is colour quenching where a coloured solution masks the emission of light from the solute. Quenching can thus be defined as a reduction of efficiency in the energy transfer process in the scintillator solution resulting in decreased light output per beta particle (28).

Various methods are used for making unquenched solutions, among them being the bubbling of argon or nitrogen through the sample and the use of ultrasonics for degassing (36) (14).

The effect of the quenching is to reduce the height of the photomultiplier output pulses, the pulse spectrum being shifted to the left as shown in figure 5.15. If we require to know the efficiency of counting in order to determine the disintegrations per minute (d.p.m.) of a sample it is necessary to employ one of the following techniques to allow for the effects of quenching.

(1) Internal standardisation. An accurately measured amount of the isotope in an unquenched form is added to a previously counted sample.

$$\text{Efficiency} = \frac{(\text{cpm of internal standard + sample}) - \text{cpm of sample}}{\text{d.p.m. of internal standard}}$$

(2) Dilution method. Involves counting a series of solutions having different concentrations of the sample. Specific activity (c.p.m./mL) is plotted against sample concentration and the graph extrapolated to zero concentration at which point the quenching would be zero.

(3) Pulse height shift or channels ratio method. The instrument is set up so that the total counting window for the isotope is split into two channels (shown in figure 5.15) and the ratio of the count in the two channels is obtained for a series of samples with a known degree of quenching. A quench correction curve is then plotted of efficiency versus channels ratio (2) (7) (8).

(4) External Standard Method. A gamma emitting source is placed close to the sample vial and compton electrons are produced from interaction of the gamma radiation with the vial and scintillator. Energy is transferred from the electrons to the solute, the energy transfer process being affected by the degree of quenching of the sample. It is therefore possible to construct a curve relating the counting efficiency of a beta emitter to the observed count rate for the external standard source (26).

**Instrumental methods of quench correction**

The first two methods of quench correction can of course be used on any liquid scintillation counter. In recent years however, automatic systems utilising the channel ratio, external standard and external standard channels ratio methods have been developed. In automatic systems using the channels ratio method, if five or six discriminators are used it is possible to set up the instrument for intermixed counting of say $H_3$ and $C_{14}$ and also for double labelled samples (9) (25). From figure 5.15 it can be seen that the total $H_3$ window is between levels L1 and L4 and the total $C_{14}$ window between levels L3 and L5. For quench correction the ratio $\frac{\text{cpm on scaler B}}{\text{cpm on scaler C}}$ is used for the $H_3$ correction curve and the ratio $\frac{\text{cpm on scaler B}}{\text{cpm on scaler A}}$ for the $C_{14}$ curve. It should be noted that the $H_3$ correction curve is accurate for both chemical and colour quenching. In the case of $C_{14}$ the curves for chemical and colour quenching diverge as the degree of quenching increases, and a separate curve should be plotted for coloured samples.

In the external standard method, a radium, barium or caesium gamma emitting source is used. For a high energy source such as radium 226 the gross external standard count rate can be used for quench correction, since, with the standardising channel correctly positioned, few of the sample counts appear in the external standard window. With lower energy sources such as caesium 137 and barium 133, net external standard counts should be used as part of the sample spectrum appears in the external standard window. In order to obtain the net external standard count rate, the sample is first of all counted and its count rate computed and stored. The external standard is then positioned and the count rate for the external standard plus sample is determined. The net count rate for the external standard is then obtained by subtraction. In order to do these calculations automatically and print out data for both sample and external standard it is necessary to build in electronic computing circuits which in the latest instruments are in the form of integrated microcircuits.

The number of compton electrons produced by the external gamma source depends on the counting geometry (sample volume, vial thickness and positioning of the source), the electron density of the solution and temperature. The dependence of the external standard count rate on sample volume necessitates the plotting of a family of quench correction curves for various volumes. This can be overcome by applying the channels ratio techniques, when it is found that except for highly quenched samples, the external standard ratio is only very slightly affected by volume. Another method of overcoming volume sensitivity is to use a compound external standard. In this method, introduced by Packard Instruments, as the sample volume varies, the change in counting geometry of a low energy Americium 241 source offsets the change in the radium 226 external standardisation count rate. In another method of automatic quench correction used by Beckman Instruments, the quenched sample is measured, then the gain of the system is adjusted when the pulse spectrum is shifted so that the end points correspond to the unquenched state. In this way automatic compensation is made for the effects of quenching.

### Electronic calculators

In addition to the calculation and storage of information required for quench correction, electronic circuits may be used for automatic subtraction of background and the calculation of statistical information.

The information contained in the scalers is fed to the computer section as is the information from the timer. The counts per minute are then calculated, the division of count data by time usually being performed by a process of progressive subtraction, the data being stored in register units composed of multivibrator circuits. The scintillation counter background was previously measured and controls on the background subtraction unit set, so that the background count rate may be subtracted from the sample count rate determined by the computer and stored in the quotient register. The count rate is computed perhaps every 1¼ secs allowing continuous digital display. There are of course many different methods used in performing calculations electronically and it is impossible to attempt a complete description here. Recently scintillation counters have been introduced which give a printout of quench corrected d.p.m. An instrument of this type is the Beckman DPM 100 which will give for single isotopes, the quench corrected DPM (accurate to about 5%) the sigma counting error, sample number, time and the external standard ratio. In the DPM 100 there are two methods for the automatic estimation of the sample disintegrations per minute. The first uses a fixed factor and can be used where all the samples have the same degree of quenching. The user must first of all determine the efficiency of counting, the per cent figure then being entered on digital dials above each data channel. The instrument then measures each sample, calculates the c.p.m., enters the dialled in efficiency factor and prints out the d.p.m. The second method uses the external standard quench correction technique, the channels ratio calibration number being determined for each sample. This number is then used to calculate the efficiency of each sample automatically. The efficiency factor is then applied directly to the counting data, allowing direct computation and printout of d.p.m.

### Computer techniques

The output from a scintillation counter may be fed directly into a computer and a typical program would be (1) read in the data, (2) compute the average and net count rates and quench correction ratios for standards, (3) perform statistical analysis to ensure data has a poisson distribution, (4) perform least squares analysis fit for efficiency v ratio and the error in efficiency v ratio for standards, (5) repeat 1 and 2 for samples and, (6) compute the d.p.m. and uncertainty. The d.p.m. is determined from the measured counts, the time and the computed efficiency.

In order that data is fed in correctly, the standards, background and samples (for each isotope) must be counted in the correct sequence. Least squares analysis involves curve fitting by considering polynomials of successively increasing values (5) (43). The computer may of course be external to the counter although instruments are available which have built in computers (as distinct from computing or calculating circuits for calculations of net counts, ratios, background subtraction)/and an instrument of this type is the Intertechnique liquid scintillation counter which has a 1024 word 12 bit integrated circuit computer with a magnetic core memory for storing quench curves. Another method recently introduced, an example of which is the Picker Nuclear DIRAC system, couples the output from the anistron 11 liquid scintillation counter to a DAC 512 In-Lab computer. This desk top computer which uses decimal language, has a 120 storage registers each of which can accept 12 characters, (9 decimal digits, sign and exponent of 10). A special interface allows the computer to be used for off line calculations independently of the liquid scintillation counter.

## SOLID SCINTILLATOR SPECTROMETERS

**Detecting crystals**

Single, dual and triple channel gamma spectrometers use similar linear amplifiers and pulse height analysers to the liquid scintillation systems just described. The detector system is of course different, and may utilise a 2" or 3" well type crystal of thallium activated sodium iodide. A typical modern instrument has between 3½ – 5" of lead shielding around the side of the detector. Steel shielding may also be employed between the detector and sample changer of automatic systems to ensure a relatively constant background while counting. Both differential and integral operations of the pulse height analyser is possible. In the differential mode, wide or narrow window widths may be chosen.

When gamma radiation is absorbed by the transparent crystal, a quantum of energy is transferred to an electron. The electron in turn loses its excess energy in interactions within the crystal, resulting in the emission of photons of light, which are detected by a photomultiplier. Sodium iodide crystals are enclosed in aluminium foil which not only keeps out moisture in the air and extraneous light, but serves as a reflector for the internal light scintillations. Very large thallium activated sodium iodide crystals are now manufactured and can be obtained in various sizes up to 16" diameter. It is necessary to use a matrix of several photomultiplier tubes in order to monitor the output from large size crystals and this technique is used in medical gamma scintillation cameras. The dimensions of the crystal which should be used for gamma spectrometry depends on the energy of the incident radiation (30). Only a part of the incident radiation leads to a total peak and when expressed as a function of the quantum energy this is known as the photofraction. Nablo and Morten have discussed the efficiency and the photofractions of thallium activated sodium iodide crystals (33).

Normally glass or aluminuim windows are used but for low background work, quartz which has a low potassium content, is utilised. The crystal should have a potassium content less than 1 part per million. Normally a 5" crystal with 6" of shielding would give a background of about 750 cpm above a bias setting of 100keV. In a low background system, this would be reduced to 600 cpm, with reductions in the photo peak due to K and RaC. Recently caesium iodide crystals activated with thallium or sodium have become available. Caesium iodide is not hygroscopic, can withstand shock and rapid changes of temperature, and sections can be machined and bent to a variety of shapes. Thallium activated caesium iodide gives a gamma ray resolution similar to a sodium iodide crystal (39). Activation of caesium iodide with sodium results in almost double the light output obtained from a thallium activated crystal and with a shorter decay time (6).

Many types of crystal and other solid scintillators are available. These include lithium (europium activated) crystals for detection of neutrons, calcium fluoride (europium activated) for X-rays and beta particles, and calcium iodide for gamma rays. Organic crystals of anthracine and stillbene are used for detection of beta or alpha particles, and neutrons or protons respectively. Plastic scintillators consisting of scintillation chemicals in polyvinyl-toluene are available, and give 65% of the light output of a similar size anthracene crystal. They have low sensitivity to gamma rays and may be used for low activity alpha particle detection. Cerium activated silicate glasses containing lithium are very efficient neutron detectors (1).

**Spectrometer circuits**

The upper level discriminator is used to slowly sweep through the pulse spectrum in a single channel pulse height analyser, the pulse height scale having first been calibrated using a source of known energy. With a single channel, only pulses falling in a certain range (defined by the counting window) are recorded and in automatic scanning spectrometers, a strip chart record of the activity and energy of the isotope is obtained. The recorder chart drive mechanism drives a potentiometer which provides the sweep voltage for the pulse height analyser. The window width controls are manually set for each scan, typical scanning times varying from 15 minutes to 96 hours.

Multi channel analysers are used so that pulses in different energy ranges can be counted simultaneously, and instruments with from 20 to over 4,000 channels are available. The electronic circuits in these instruments are somewhat complex but they comprise the following basic elements.

(1) An analogue to digital converter, the amplitude of the input pulse being associated with a specific energy channel.
(2) A memory store.
(3) A data display system that enables the operator to monitor the information stored in the memory.
(4) An output device such as a printer, recorder or punch tape or card.

Many types of analogue to digital circuits have been described. These include multi discriminator, and pulse height to time converters. In the latter method, the input pulse is applied to a pulse stretching circuit, or to a capacitor which is then discharged at constant current so that the capacitor voltage decays linearly to zero. Pulses from a precision clock oscillator are allowed to pass to the accumulating and storage devices by a gating circuit, which opens when discharge starts and closes when the capacitor voltage is zero. In the pulse stretching circuit, the length of the pulse is made proportional to the signal amplitude, and a similar gating technique may be used to pass time pulses to the storage circuit. The number of oscillator pulses registered is a measure of the input signal amplitude. Another method used, compares the input voltage with internally generated reference voltages, using a register circuit composed of flip flops. If the signal voltage is less than the initial reference voltage, the first flip flop is set to the 0 state and if it is greater to the 1 state. A second reference voltage is then applied, and the second flip flop is set to a state determined by the condition of the first flip flop. This process is repeated and at each step the binary number in the register becomes closer to the amplitude of the input signal. When all the reference voltage steps are completed the appropriate memory store (connected to the register) holds the signal voltage. The input signal voltage may also be used to control the frequency of an oscillator, the number of oscillator pulses being fed into the memory store. Memory devices used include delay lines, magnetic cores and cathode ray storage tubes. Space prohibits a detailed discussion of the circuits employed, a good description of which has been given by Chase (13).

In conclusion, mention should be made of semiconductor detector systems used for gamma spectrometry and X-ray fluorescence analysis. Lithium drifted silicon and germanium detectors compare favourably with conventional proportional counters up to energies of 5MeV and give perhaps 10 times the resolution. Using side entry, better than 2% resolution is obtained for gamma rays up to 11 MeV. In the Nuclear Enterprise high resolution spectrometer, the semiconductor detector output is amplified by a charge sensitive pre-amplifier, which utilises two field-effect transistors in parallel at the input stage to provide low noise performance. The field effect transistors may be mounted within a cryostat chamber, when cooling and the reduction of stray capacitances lowers the noise level by more than 30%. The charge sensitive input circuit makes the signal amplitude independent of variations in detector capacitance. The charge sensitivity is 50mV/MeV loss in a silicon detector. A conventional high stability linear amplifier and pulse height analyser are used after the pre-amplifier unit.

Combined amplifier, pulse height analysing, scaling, timing, ratemeter and high voltage supply circuits may be combined in one compact unit by the use of integrated circuit technique. Such an instrument is the Nuclear Enterprise NE8622 integrated circuit spectrometer, which with external pre-amplifier units may be used for scintillation, proportional, geiger and semiconductor counting.

**Activation analysis**

In activation analysis, an element is converted into a radio active isotope by bombardment with neutrons, high energy protons, or alpha particles. The intensity of the radiation emitted by the sample is a measure of the amount of the element originally present. Neutron tubes in which ions from an ion source are accelerated by a 125kV potential and impinge on a tritium cathode liberating neutrons are a recent development in this field. The detection system may utilise a geiger counter or a crystal scintillation counter, which is used in conjunction with a single or multichannel pulse height analyser.

**Mössbauer spectrometry**

In the Mössbauer spectrometer a source is moved relative to an absorber and a change in energy of the emitted gamma ray is produced by making use of the Doppler effect. In this way, the resonance spectrum of an absorber specimen containing a Mössbauer nuclide is scanned. In order to obtain nuclear resonance absorption the emitting atoms are placed in a crystal lattice so that the mass of the recoiling structures is extremely large and recoil-free emission is obtained. Very high resolution can be acheived and information obtained on the structure of compounds.

The spectrometer may consist of the source, driven by a vibrator unit with an electronic servo amplifier control unit, and a scintillation or proportional detector feeding an amplifier and single channel pulse height analyser. The output from this analyser is passed to a multichannel analyser which stores and displays the spectrum. A crystal oscillator is used in a waveform generator which provides a signal, via a velocity attenuator, to the servo amplifier and also triggering pulses for the multichannel analyser. In the servo amplifier the signal from the velocity attenuator is compared with a signal from a velocity coil in the vibrator, any difference being amplified and used to modify the signal applied to the driving coil. Velocities of up to 60 cm sec$^{-1}$ can be obtained.

### References

1. Anderson, D. G. et al. Instr. & Measurements Conf. Stockholm. Vol 2. 616. Academic Press.
2. Baillie, L. A. (1960) *Int. J. Appl. Rad. and Isotopes.* 8, 1.
3. Baker, S. C. (1959) Trans. I.R.E. NS-6. 57.
4. Bernstein, W., Jerknes, C. B. and Steele, R. in Bell and Hayes Liquid Scintillation Counting. Pergamon.
5. Blanchard, F. A. (1963) *Int. J. Appl. Rad. and Isotopes.* 14, 213.
6. Brinkman, P. (1965) *Physics Letters.* 15. 305.
7. Bruno, G. A. and Christian, J. E. (1961) *Anal. Chem.* 33, 650.
8. Bush, E. T. (1963) *Anal. Chem.* 35, 1024.
9. Bush, E. T. (1962) Nuclear Chigago Technical Bulletin No. 13.
10. Cameron, J. F. and Boyce, I. S. (1962) in Tritium in the Physical and Biological Sciences. I.A.E.A.
11. Chase, R. L. (1961) Linear Amplifiers in Nuclear Pulse Spectrometry. McGraw-Hill.
12. Chase, R. L. and Svelto, V. (1961) Trans. I.R.E. NS-8. 45.
13. Chase, R. L. Nuclear Pulse Spectrometry. McGraw-Hill.
14. Chleck, D. and Ziegler, C. (1957) *Rev. Sci. Inst.* 28, 466.
15. Dance, J. B. (1967) Electronic Counting Circuits. Iliffe.
16. Dean, K. J. (1965) An Introduction to Counting Techniques and Transistor Circuit Logic. Champan and Hall.
17. Doyle, J. M. (1963) Pulse Fundamentals. Prentice Hall.
18. Emmer, T. L. (1962) Trans. I.R.E. NS-9. No. 3.
19. Ekco Electronics. Technical Bulletin S19. (1961).
20. Ekco Electronics. Technical Bulletin S4. (1958).
21. Fairstein, E. (1956) *Rev. Sci. Inst.* 27. 476.
22. Fairstein, E. (1962) Pulse Amplification in A. H. Snell (ed). *Nuclear Instruments and their uses.* Wiley.
23. Frank, R. B. (1962) Trans. I.R.E. NS-9. 345.
24. Frank, R. B. (1962) Trans. I.R.E. NS-9. 345.
25. Hendler, R. W. (1964) *Anal. Biochem.* 7, 110.
26. Higashimura, T. (1962) *Int. J. Appl. Rad. and Isotopes.* 13, 308.
27. Kallman, H. and Furst, M. (1958) in Bell and Hayes. Liquid Scintillation Counting. Pergamon.
28. Kallmann, H. and Furst, M. (1958) in Bell and Hayes. Liquid Scintillation Counting. Pergamon.
29. Klein, P. D. and Eisler, W. J. (1966) *Anal. Chem.* 38, 11. 1453.
30. Kreger, W. E. and Brown R., M. (1961) Nuclear *Instr. and Methods.* 11, 290.
31. Lloyd, R. A., Ellis, S. C. and Hallows, K. H. (1962) Tritium in the Physical and Biological Sciences. I.A.E.A.
32. McDonald, D. F., Dunn, B. J. and Braddock, J. V. (1960) Trans. I.R.E. NS-7. 17.
33. Nablo, S. V. and Martin, T. C. (1961) *Int. J. Appl. Rad. and Isotopes.* 10, 55.

34. Nuclear Chigago publication (1963) No. 711580. Liquid Scintillation Counting.
35. Nuclear Chigago publication (1963) No. 711580. Liquid Scintillation Counting.
36. Ott, D. G. (1955) *Nucleonics.* 13. **5,** 62.
37. Packard, L. (1958) in Bell and Hayes Liquid Scintillation Counting. Pergamon.
38. Rapkin, E. (1964) *Int. J. Appl. Rad and Isotopes.* **15,** 2.
39. Schmidt, C. T. (1960) Scintillation Counter Symposium. Washington. U.S.A.
40. Sharpe, J. (1961) *Electronic Technology.* June.
41. Sharpe, J. and Thompson, E. E. (1958) Photomultiplier Tubes and Liquid Scintillation Counters. Atoms for Peace Conference. Geneva.
42. Sharpe, J. and Stanley, V. A. (1962) in Tritium in the Physical and Biological Sciences. I.A.E.A.
43. Spratt, J. L. (1965) *Int. J. Appl. Rad. and Isotopes.* **16,** 439.
44. Stanley, V. A. (1966) Trans. I.E.E.E. 10th Symposium on Scintillation Counting.
45. Swank, R. B. (1958) In Bell and Hayes Liquid Scintillation Counting. Pergamon.
46. Swank, R., Phillips, H., Buck, W. and Basile L. (1958) Trans. I.R.E. NS-5. 183.
47. Tamers, M. A. (1964) *Packard Inst.* Technical Bulletin. No. 12.
48. Wang, C. H. and Willis, D. L. (1965) Radio Tracer Methodology in Biological Sciences, Prentice Hall.
49. Watt, D. E. and Ramsden, D. (1964) in High Sensitivity Counting Techniques. Pergamon.
50. Whetstone, A. L. (1964) Trans. I.R.E. NS-9. 345.

# 6 Electronics in Analytical Separation Techniques

## GAS CHROMATOGRAPHY

**Principle**

In gas chromatography, the sample is distributed between the moving phase and a stationary bed of liquid or solid contained in a column. The columns employed may be of the packed or capillary type, and some of the sample components in the gas phase are retained by the column materials longer than others. In packed columns, a granular material is used, which is an absorbent for gas-solid chromatography, and an inert material supporting the liquid in gas liquid chromatography. Capillary columns are coils of narrow bore tubing coated internally with the liquid or solid phase. Various types of detectors are used for monitoring the column effluent, the sample components emerging in the reverse order of their retention by the column. The column is maintained at a constant temperature, or for programmed temperature work the temperature varied in a definite way during analysis. The coupling of a gas chromotagraph to a mass spectrometer and infra-red spectrophotometer has been described in chapter 4.

Thermal conductivity cells, gas density cells, flame ionisation and radiation detectors are used for monitoring the column.

**Radioactive detectors**

Detectors using a radioactive source (e.g. strontium 90) for gas ionisation have already been referred to briefly in chapter 5. An example of this type of equipment is the Pye argon chromatograph, in which the absorption of ionising radiation by the gases entering the ionisation chamber, produces ions and excited non-ionised (metastable) atoms and molecules. Argon is used as the carrier gas and the metastable atoms decay to the normal state only when pure gas is present. Small traces of other gases cause the metastable atoms to transfer energy by collision, and the gas molecules being carried into the chamber by the argon carrier are ionised. The collector electrode is placed inside the chamber, and the ionisation current produced (about $10^{-10}$ A) flows through an external high ohmic value resistance. The voltage developed is amplified and fed to a recorder, the standing current due to the argon carrier alone being backed off (12).

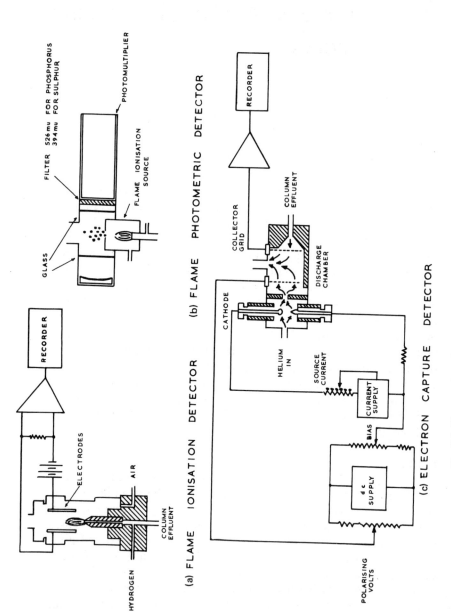

*Fig. 6.1.* Flame and spark discharge ionisation detectors.

Radio chromatography equipment may also be used with the argon chromatograph for $C_{14}$ labelled compounds. A normal chromatograph is obtained from the argon ionisation detector, but the major part of the column effluent is diverted through a side arm where it is heated and passed through a combustion tube. The gas stream from this tube passes to a proportional counting chamber, a small amount of $CO_2$ being added to give suitable counting characteristics. When a molecule derived from the labelled compound occurs, beta particle emission produces ions by gas multiplication, and a current pulse results. This is amplified and fed to an Ecko N701A ratemeter, comprising a variable gain amplifier, discriminator, pulse shaping circuits and a diode pump ratemeter. Other radioactive sources used in gas chromatography systems are tritium, radium 226 and americium (21).

**Flame ionisation detectors**

The flame ionisation detector is shown in figure 6.1(a). The column effluent is filtered and mixed with hydrogen at the base of the jet. The gases are ignited by a platinum wire igniter, oxygen or air being fed axially round the base of the jet. The current from the detector (about $10^{-12}$ A) flows through external resistances and is proportional to the number of ions or electrons formed in the flame gases. The detector responds to traces of organic compounds and an electrometer amplifier is used to amplify the signal to 1–10mV (suitable for a potentiometric recorder).

**Electron capture detector**

An electron capture detector, in which ions and electrons are produced by a d.c. discharge through helium, is shown as figure 6.1(c). Electrons from the discharge chamber enter the detection chamber and are attracted to the collector giving rise to a standing current. When the column effluent, containing electron absorbing molecules arrives, the current level falls and the drop in current is related to the concentration of the sample. The background current is controlled by the level of the polarising grid potential. Carbon dioxide is usually added as a diluent which absorbs photons produced by the discharge so that the detector response is practically all due to electron capture. Full wave rectifiers with smoothing and automatic voltage regulating circuits are used in the power supply units. As the input impedance of the electrometer amplifier is high it is important that the detector elements and insulators are not contaminated. This would result in shunting effects and reduce detector efficiency. A revue of ionisation detectors has been given by Thorne (26). A flame ionisation detector which can be constructed from readily available materials has been described by Hughes. A pH meter is used for the electrometer amplifier (7)

## Thermal conductivity detectors

Four hot wire filaments (or thermistors) mounted in separate cavities in a metal block, and arranged as a bridge circuit are used in the thermal conductivity detector. The carrier gas flows over two of the elements, while the carrier diluted with the sample gas (i.e. the column effluent) flows over the remaining two. The temperature gradient produced, depends on the thermal conductivity of the gas and the wire filament temperature. Initially as

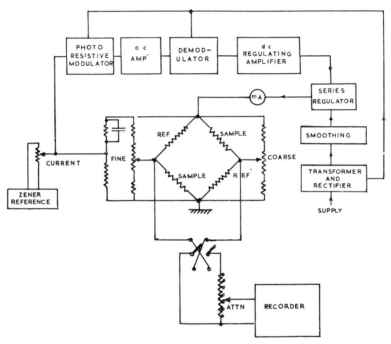

*Fig. 6.2.* Beckman Thermal conductivity detector circuit.

the bridge is balanced with carrier gas flowing over the wire filaments, a state of temperature equilibrium is reached, the wire resistances then having a certain value. When the reference gas, diluted with sample, flows over one pair of wire filaments their resistance changes and the bridge circuit is unbalanced. The output of the bridge is passed through an attenuator network to a recorder. The sensitivity of this type of detector necessitates a stabilised d.c. supply to the bridge circuit (better than ± 0.02%).

A simplified diagram of the electronic control circuit used on the Beckman thermal conductivity detector is shown in figure 6.2. The bridge circuit is first balanced by the coarse and fine balance potentiometers so that the recorder reads zero with the carrier (reference) gas flowing. Part of the voltage applied to the bridge is compared with a reference voltage and the error signal converted to a pulsating form by means of a half wave photoresistor-neon lamp modulator. The signal is then passed through an a.c. amplifier, stabilised by both d.c. and a.c. feedback loops, to a two transistor demodulator. The resulting d.c. signal is then applied to the regulating amplifier, which controls the series regulating transistor, so that the bridge voltage is maintained at the correct level. A current control, which sets the voltage applied to the bridge, forms part of the reference voltage-bridge signal comparison circuit. It is set so that the bridge current (indicated on a milliammeter) is at the value required for optimum performance. In hot wire detectors the chromatograph peak area is inversely proportional to the gas flow rate and proportional to the cube of the bridge current (19).

## Gas density detectors

In the gas density detector, the carrier gas is split into two balanced streams and two thermal elements, (hot wires or thermistors) which form two arms of an electrical bridge, detect any difference between the streams. The effluent from the column is also split in the detector, and when sample gas molecules heavier or lighter than the carrier occur, the flow of gas over the thermal elements is unbalanced (14) (13). The column effluent enters the detector at a different port to the carrier gas and does not flow over the elements so there is no possibility of contamination.

## Ultrasonic detectors

The ultrasonic detection system developed by the Micro Tek subsidiary of Tracor, Inc., is universal and non-specific (15) (30). The detector cell uses four piezo-electric crystal transducers in pairs. The two transmitting transducers are connected to a 6MHz oscillator and the two receiving transducers to the sensitive phase meters as shown in figure 6.3.

The speed of sound depends upon the specific environment or medium through which it is passing and the presence of any vapour or gas will cause a change in that speed. If carrier gas is passed through both sides of the Micro Tek ultrasonic cell the appearance of a sample on one side will change the speed of sound relative to the other. This relative change is detected as a phase change in the sound wave which is recorded graphically after suitable signal processing in the measuring circuits.

Thermal conductivity detectors are affected by temperature, pressure, flow and have electrical limitations. These disadvantages are overcome in the ultrasonic detector as the temperature coefficient of the speed of sound is one tenth of the temperature coefficient of a thermal conductivity cell and since the speed of sound equals $K\sqrt{\frac{P}{D}}$ where P equals pressure, K equals a constant and D equals the density of the gas, any change of pressure is compensated by a resulting change in density so that the detector

is relatively insensitive to flow and pressure. The detector also has ten to thirty times the sensitivity of a thermal conductivity cell depending upon the carrier gas and molecular weight of the unknown. It also has a faster response although on the other hand it costs three times as much and requires a stable oscillator.
It is also important to have sufficient back pressure in order to maintain gas density in the cell to give a reasonable signal level without noise. The gas pressures required for some carrier gases are as follows: 30 lbs for helium, 80 lbs for hydrogen and 50 lbs for nitrogen.

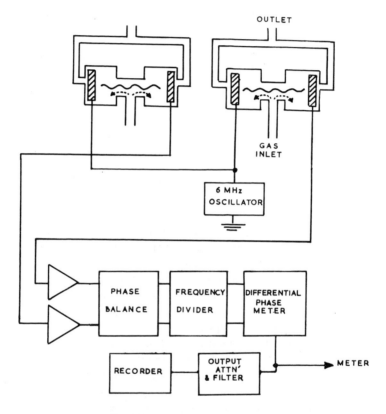

*Fig. 6.3.* Ultrasonic detector operation.

## The micro coulometer

The Dohrmann micro coulometric titration system is a specific detector, in which the sample is converted to a titratable form by a reduction or oxidation furnace, and passed to a coulometric cell (silver or iodine). Electric equipment consist of an integrating amplifier and recorder so that coulombs can be determined. Using Faraday's laws of electrolysis we can then determine how much of the sample is present. The system is a highly sensitive absolute detector (one nanogram is the minimum detectable quantity).

### Flame photometric detectors

Another specific detector is the flame photometric type, used for phosphorus and sulphur containing compounds. A hydrogen ridge flame is used to combust the sample and the excited molecules emit light which is detected by a photomultiplier. The arrangement is shown in figure 6.1(b). No light from the sample is seen by the photomultiplier as the flame is in a cup. The detector is some 2,000 times more specific for phosphorus than for hydrocarbon, and has a sensitivity of 1 nanogram. Bands of wave lengths are produced and there is in fact about 10% overlap between phosphorus and sulphur.

### The Coulson electrolytic detector

In this system a pyrolyser, a gas-liquid contactor, a gas liquid separator and a pair of platinum electrodes connected in a d.c. bridge circuit are used. The pyrolyser converts organically bound halogen, sulphur, or nitrogen to oxidised or reduced substances that form electrolytes when dissolved in water. The electrolytes are detected by the change they cause in the conductivity of water in the detector cell. The output from the bridge circuit is recorded on a potentiometric recorder. The detector has good selectivity and sensitivity.

### Gas chromatography systems

Dual column, multiple detector, gas chromatography systems are available in which the effluent from each column is split, so feeding two different detectors. A total of four detectors are thus available and a commercial example of this type is the F and M 402 chromatograph. In dual channel electrometer systems, two detectors can be operated simultaneously, or both outputs from a dual flame ionisation detector measured.

The amplifier system used for flame, electron capture and cross section detectors is of the d.c. or vibrating capacitor electrometer type. Sensitivities to about $4 \times 10^{-12}$ A full scale are usual. Noise is generally less than about 1% and may be as low as $2 \times 10^{-14}$ A. Background suppression by means of helical potentiometers of up to about 10 times $10^{-9}$ A is also provided.

Parametric amplifiers may be used in place of the conventional electrometer and overcome the problems of warmup delays, ageing, drift, humidity etc. Perkin Elmer use a varactor bridge (parametric) amplifier in their latest gas chromatograph and the circuit gives a maximum sensitivity of $5 \times 10^{-12}$ amps with $5 \times 10^{-14}$ amps long term drift, and $7.5 \times 10^{-15}/°C$ drift with temperature. A simplified circuit is shown as figure 6.4.

Varactor diodes act as voltage variable capacitors, and when an input signal is applied to the bridge circuit the varactor capacitances alter and an unbalanced condition results. Part of the

| Type | Sensitivity | Performance |
|---|---|---|
| Thermal Conductivity cell | Above about 100°C filaments most sensitive Below about 100°C thermistors more sensitive | Time constant of order of seconds. Most universally applicable and widely used. Wide dynamic range. |
| Gas Density Detector | Sensitivity and limit of detection of same order as Thermal Conductivity cell. | Time constant of order of seconds. |
| Flame Ionisation | Approximately 1000 times sensitivity of Thermal Conductivity Detector. | High sensitivity and low internal volume (small sample requirement). Sample must burn or ionise in flame. Limited dynamic range. Time constant of order of milliseconds. |
| Radiation Detector (argon carrier) | Approximately 300 times sensitivity of Thermal Conductivity Detector. | Compares favourably with flame ionisation. Sample must be ionised by argon in its metastable state. Good dynamic range. Time constant of order of milliseconds. |
| D.C. Discharge | Approximately 100 times sensitivity of Thermal Conductivity Detector. | Low dynamic range. |
| R.F. Discharge | Approximately 200-300 times sensitivity of Thermal Conductivity Detector. | Low dynamic range. |
| Ultrasonic | Approximately 10-30 times sensitivity of Thermal Conductivity Detector. | Extremely wide dynamic range. Time constant of order of milliseconds. |

TABLE 6.1.
*CHARACTERISTICS OF GAS CHROMATOGRAPH DETECTORS*

10MHz carrier oscillator output (depending on the degree of unbalance) is then fed to the a.c. amplifier, the output from which is phase sensitive demodulated and d.c. amplified. In the gas chromatograph application the parametric amplifier and associated high ohmic value feedback resistors are mounted in a temperature controlled enclosure (22) (5).

A recent development is controlled sub ambient chromatography. Equipment of this type is available from Microtek Instruments Inc., and has a temperature range of $-180°C$ (using liquid nitrogen) to $+400°C$. A solenoid valve controls the coolant injection

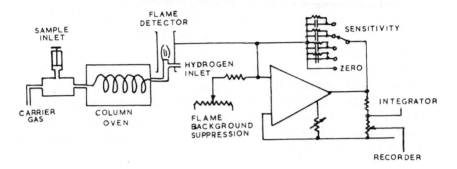

PERKIN ELMER GAS CHROMATOGRAPH PARAMETRIC AMPLIFIER SYSTEM

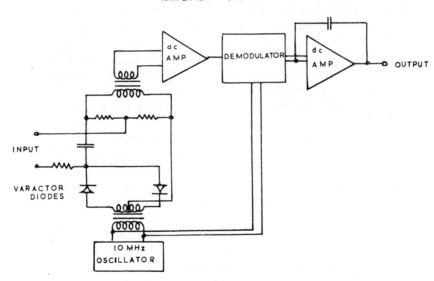

*Fig. 6.4.* Parametric amplifier detector.

into the chamber of the column oven, and is operated by a variable pulse time proportional cryogenic controller. A temperature programmer with a proportional silicon controlled rectifier circuit controls the oven heaters. Thermocouple temperature detectors are used for both cooling and heating systems, and a cross over network is coupled to the temperature programmer for a smooth changeover from sub ambient temperatures during programming.

**Baseline drift correction**

Dual column instruments in which the carrier flows through one column and the sample plus carrier through the other, reduce the problem of an unstable base line. In practice it is impossible to pack two columns absolutely identically and better results may be obtained using electronic circuits to automatically correct for base line drift. In modern instruments such as the Varian Aerograph 475,476 digital integrator, automatic baseline correction occurs when the peak is not being integrated. During the time when the peak is integrated, the baseline is maintained at the same corrected level as just prior to the peak. Baseline correction may be applied automatically or manually. In the automatic mode, the correction range is from + 1000 to − 500 uV at rates selectable from ± 0.05 to 50 uV per second. A fast rate of 60 uV per second is also provided. The operation of a digital integrator is outlined in the section on Amino Acid analysis.

**Integrators**

The principle of electronic integration was outlined in chapter one and electronic integrators are widely used for measuring peak areas in gas and liquid chromatography. Integrating amplifiers are available in which the output is fed to a recorder, the height of the steps produced on the trace when peaks occur representing the peak area. Digitising and print out facilities are available and these items together with the amplifier may be incorporated in one compact unit. Digital integrator units usually incorporate a voltage to frequency converter, the pulses produced being fed to an electronic counter. The counter is controlled by a gating circuit which is opened to allow pulses to pass through when a peak occurs. The counter information can be transferred to a memory register which in turn feeds the printer. Accuracy or reproducibility of peak areas is of the order ± 0.1%. As in other branches of analytical chemistry, data processing techniques may be applied. The outputs from various chromatographs can be recorded on tape recorders then subsequently played back and passed through an integrator and data processor to a teletype printer. Both on line and off line systems of this type are available from Infotronics Corporation.

## LIQUID CHROMATOGRAPHY

In liquid chromatography, the associated electronic equipment consists of (1) A U.V. column monitor and (2) A fraction collecting system. The complete equipment comprises a container for the elution liquid, which may be supplied at pressure with a pump, a column containing the absorption bed and the analysing and fraction collecting apparatus.

## Column monitors

In the simplest column analyser, the effluent passing through a flow cell is monitored at a fixed wavelength. The wavelength chosen is usually 2537 Å as proteins, some peptides, steroids, amino acids, nucleotides and enzymes absorb at this wavelength. An instrument of this type is the L.K.B. Uvicord I, and the 2537 Å emission line from a low pressure mercury lamp is isolated by liquid and black glass filters. The liquid filter is located between the light source and the quartz sample cell, and the black glass filter between the cell and the R.C.A. 935 phototube detector. The lamp supply is stabilised by a constant voltage transformer and a small heating element may be fixed to the lamp for cold room operation (i.e. between 0 and 15°C). The heater is not used for operation over the normal ambient temperature range of 15 to 30°C. The photocell current flows through a high resistance, is amplified and passed to a cathode follower output stage. This stage may feed a 10 mV, 100 mV potentiometric recorder or a 0.3 mA moving coil recorder. In the Uvicord II (figure 6.5) a fluorescent rod may be inserted in the light path to give an additional wavelength of 2,800Å. This is useful for proteins and their derivatives whose U.V. absorbance is greatest at this wavelength.

Some instruments are designed for base line compensation and two flow cell cuvettes are used. One known as a reference cuvette monitors the input to the top of the column and the other sample cuvette is used to monitor the column effluent. The light from the U.V. source is split into two paths and after passing through the cuvette is measured by two photocells. Their outputs are fed to a ratio circuit the output of which represents the quotient of the input signals and this value is plotted on a recorder. The solvent flows through the reference cuvette and the solvent plus effluent through the sample cuvette, the ratio dividing circuit typically giving a base line stability of ± 0.5% over twenty-four hours. The advantage claimed for this method over subtraction circuits is that the system sensitivity is maintained as solvent absorption increases, Monochromators may be used to set the incident radiation exactly at the absorption peak of the sample. An example is the Canalco analyser.

Instruments are available which record absorbance instead of transmission and the effect of base line shift on the trace is then very small, as a change in the transmission of light due to the solvent absorption changing has little effect on an optical density

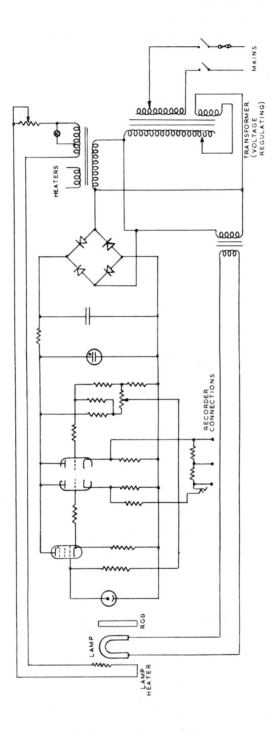

*Fig. 6.5.* The L.K.B. Uvicord II.

scale. The analyser may actuate a fraction collector at the beginning and end of absorbance peaks so that each fraction is in a separate collecting tube. An instrument of this type is the Isco UA absorbance recording analyser which also has a very narrow band width of less than 1 m$\mu$ at a wavelength of 254 m$\mu$.

Another method of relating the fractions to the recorded peak in systems where a fixed volume or a number of drops are collected, is to use an event marker which puts a 'blip' on the recorder trace whenever the fraction collector tube is changed. In the L.K.B. system two coloured ribbons are used in the chopper bar recorder, and the colour is changed whenever the fraction collector tube is changed. In this way, the analyser recorder trace alternates between red and black as the tubes change.

**Fraction collectors**

Very many methods are used for determining the amount of liquid to be collected in the fraction collector tube. They include photoelectric devices, time, syphon (volume) and contact making systems. In the latter method, the drop as it forms at the end of a dropping tube fed from the column makes an electrical circuit. Another volume sensitive system uses electrical contacts which are made when the liquid rises to the required level. A review of some early fraction collector designs has been given in Analytical Chemistry (1).

Photo sensitive resistors, diodes, transistors or photocells may be used as detectors in drop counting fraction collectors. The drop when it passes between the detector and the small light source gives rise to a change in the photo sensitive detector current. This is amplified and used to actuate a relay feeding a counting device which is preset to the required number of drops. On reaching the set number, the collector turn table is operated and a fresh tube placed in position under the column.

When using solid state detectors, problems may arise when operating in hot or cold rooms (e.g. at temperatures of 37°C and 4°C.). The photo transistor is placed in a housing with a small hole to admit light from the light source. The temperature of the photo transistor itself may therefore be considerably in excess of the ambient air temperature. This means that when operating in a hot room, the transistor leakage or dark current may approach the value of the current flowing under light conditions. The current change produced when a drop interrupts the light beam under these conditions is small, and may be insufficient to operate a relay, thus resulting in unreliable operation.

By using a suitable thermistor in the base circuit of the transistor, automatic temperature compensation may be obtained. Another method is to use a Schmitt trigger circuit (figure 6.6), which gives a constant amplitude output to the relay for varying

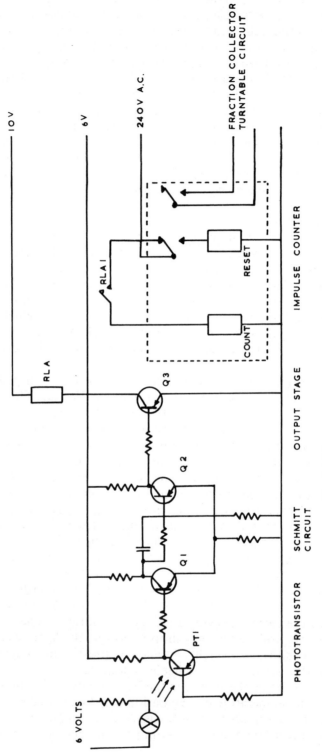

Fig. 6.6. Photo-Schmitt trigger circuit.

size pulses from the photo transistor. The relay can be used to
operate an impulse counter which is preset to the required number
of drops. The counter counts down to zero, at which point the
collecting tube is changed, and the counter automatically resets
to the preset number. The tube on the collector may also be
changed at selected time intervals, the selected collection time
being related to the chart speed of the analyser recorder, so
enabling easy identification of the tube in which a selected peak
is collected (18). Many fraction collectors use circular turntables, but recently rectangular units have been introduced composed of a number of racks, each holding perhaps ten tubes which
are moved in sequence under the detector head.

### The Pye liquid chromatograph system

A recent development has been the use of the high sensitivity
argon detector in liquid chromatography. In the Pye liquid
chromatograph system, a moving stainless steel wire is used which
collects the eluent as the wire passes through a slit in a hypoder-
mic needle attached to the column. The wire is first cleaned in an
oven under a stream of argon and then coated with the eluent as
described above. The wire next passes to a second oven, where
the solvent is removed by heating, and then to a third oven where
the materials are removed from the wire by heating to 700°C. The
pyrolysis products are then carried to an argon ionisation detector
by a stream of argon. An evaluation of this method has been given
by Cropper and Heinekey and both moving wire and chain systems
have been described (3) (8) (24).

### Amino acid analysers

In automatic chemical analysis systems, such as amino acid
analysers, electronic circuits can be used not only for voltage
stabilisation, temperature control and multichannel recorders but
also for base line correction, integration and digital readout. In
the amino acid analyser, the various amino acids are first separated
in ion exchange chromatography columns and then reacted with
ninhydrin. The reaction produces coloured compounds which are
measured by colorimeters at 570 m$\mu$ and 440 m$\mu$. The colorimeter photocell output may be fed into a digital integrator unit
such as the Beckman model 125, which not only gives a print out
of peak area in digital form but also performs automatic base line
correction. A simplified diagram of this integrator circuit is shown
as figure 6.7.

The standard 570 and alternate 440 m$\mu$ signals are amplified to
ten volts and applied via a switching circuit to a logarithmic/inte-
grator converter circuit. A parallel circuit also feeds a multi channel
recorder via a potential divider network (the third 570 m$\mu$ colorimeter output is not fed through the integrator). The ten volt

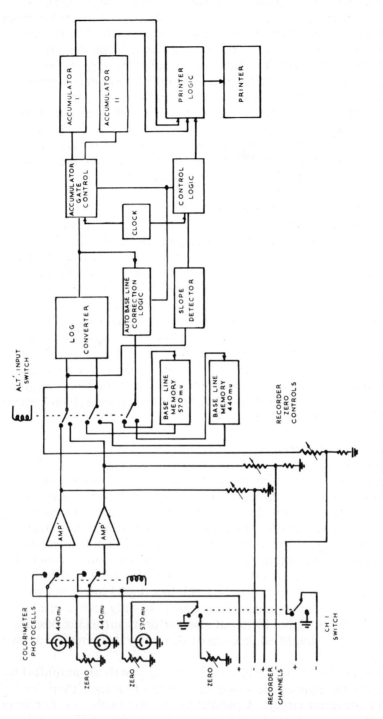

Fig. 6.7. The Beckman 125 digital integrator.

signal level represents zero optical density and the colorimeter signal that does not contain a peak is the analyser base line. The analyser signal and the output of either the 570 or 440 base line memory circuits are fed to the logarithmic converter. If any difference exists between these two outputs, an automatic base line converter circuit adjusts the internal base line of the integrator unit until it matches the analyser base line, and the "difference" output of the logarithmic converter is zero. The logarithmic converter/integrator circuit changes the input peak signal voltage to a train of pulses, and the number of pulses per second is proportional to the absorbance measured in the colorimeter.

When a peak begins, the slope detector circuit feeds a signal into the control logic. The peak amplitude is then monitored every second and the resulting pulses stored in an accumulator circuit. The slope detector also senses the trailing edge of the peak, and a signal from the control logic circuit closes the accumulator gate control. The control logic then commands the printer to print out the pulses stored in the accumulator. The 570 m$\mu$ input is processed, except during a period which can be preset on a timer, when the 440 m$\mu$ input is fed to the integrator. A variable base line delay time (1–100 secs) determines the length of time between the end of integration and the start of base line correction. A plateau time switch (1–100 secs) determines the length of time that the circuit will integrate a plateau occurring in the chromatograph peak. Both the base line correction rate and the slope detector circuit sensitivity can be varied between 0.0002 and 0.02 OD/minute.

The digital integrator has an accuracy of ± 1.5% of area at 25°C ± 3°C with a peak height of 2.0 OD. On normal range, a 0.1 OD peak height gives an area accuracy of ± 2.7% at 25°C. An expanded range is however also available and using this, the 0.1 OD peak has an area accuracy of ± 0.5%.

## ELECTROPHORESIS

**Principle**

In electrophoresis, a charged ion migrates towards one electrode when placed in an electric field. In zone electrophoresis, a mixture to be separated is applied as a spot on a paper (Whatman No. 1 or 3 filter paper) wetted with a buffer. After electrophoresis the substances are located as a number of separate discrete bands or zones.

Many designs for electrophoresis have been described in the literature (11) (16) (25) and one arrangement is that in which the paper is immersed in a water immiscible volatile organic cooling liquid. Constant current or constant voltage power supplies are used to pass current through the paper. The buffer solution keeps the paper wet to maintain current flow, and also maintains pH during the experiment so that the substances will migrate under fairly constant conditions. With a constant voltage supply, as the paper warms up its resistance falls and more current flows giving rise to further heating. Cooling is therefore, important, as with a stable temperature, the conductivity and ionic mobility are more constant. To prevent the tendency for the temperature to rise in a constant voltage system, the voltage may be reduced as the paper warms up, so that there is a lower power dissipation.

**Power units**

In low voltage power supply units the output may be 500 volts at 50mA, but for voltage gradients in excess of 20–30 V/cm high voltage supplies up to 10,000 volts are utilised. Since the current supply at this voltage may be 0.5 or 1 ampere, care must be taken in their use. The heat dissipated may be removed by passing cooling water through glass coils inside the electrophoresis tanks. The tanks containing the paper and white spirit or toluene cooling liquid, should be enclosed in a chamber similar to a fume cupboard with a sliding door. A flame proof microswitch should be installed in the chamber to automatically cut off the E.H.T. voltage when the door is open.

An automatic $CO_2$ fire extinguisher system should also be incorporated and for safety, a flame proof thermostat may be placed in the tank to cut off the power before the temperature of the liquid rises near to its flash point. In addition a flow or pressure switch can be used in the cooling water supply as a safety interlock.

Large power units usually use a three phase variac to control the voltage fed to a three phase transformer, the output of which is at high voltage. The transformer output may be fed to a bridge rectifier circuit and smoothed before being metered and passed into the electrophoresis chamber. Earth leakage and current overload relays are also incorporated.

**Materials**

In addition to paper, other supporting materials have been found. These include membranes and gels (polyacrylamide, agar, starch gel), powders (glass powder or beads), and density gradients (Ethanol and Sucrose). A disadvantage of paper is that substances may be bound to it by adsorption. Gel techniques have a higher separating power than paper, due to the gel pores acting as molecular sieves, although some methods are technically more difficult than paper. The polyacrylamide gel is becoming increasingly popular due to the simplicity of the technique and its superior properties.

## Continuous flow electrophoresis

In continuous flow electrophoresis (used for preparative purposes), the buffer flows slowly and uniformly over a vertical paper curtain. The solution to be separated is applied continuously at a point on the upper edge of the paper. A horizontal electric field is applied, and at the bottom there are a number of drip points so that separated components are collected in fraction collection tubes.

## Moving boundary electrophoresis

Moving boundary electrophoresis takes place in a free liquid medium. A U-shaped Tiseluis cell of rectangular cross section (and placed in a controlled temperature bath) is used. It is partly filled with the sample solution which is overlaid with a buffer electrolyte. A constant current power supply is applied (e.g. 300 V at 20 mA) and ions from the solution tend to move into the buffer with a velocity that depends on their charge, size and shape. The boundary between the buffer and solution is observed using a Schlieren or Raleigh optical system. Changes in refractive index, due to the boundary between the solution and buffer being split up by the migrating ions, are measured.

## Electro-focusing columns

An electro-focusing column has recently been introduced by LKB for the separation of proteins. A density gradient non ionic compound, in which carrier ampholytes are dissolved, is established in the column. The sample is introduced by layering or distribution throughout the column and when a voltage is applied over the mixture (up to 1,200 V) the carrier ampholytes form a pH gradient. The proteins in the sample then migrate until they reach the point when they are electrically uncharged. A complete experiment may take some time (up to three days) after which the column can be emptied without remixing of the separated protein (28).

# ELECTRONIC CONTROL OF ULTRACENTRIFUGES

The ultracentrifuge is used for both preparative and analytical work, electronic circuits being utilised for speed and temperature control.

The electric motor is series wound, that is the field winding is in series with the armature. This type of motor is used as it gives a fast acceleration, a wide speed control range and a high starting torque. The motor drive shaft is connected through a step up gear box to the rotor drive. The centrifuge speed, (at present a maximum of 65,000 rpm), is controlled by varying the voltage

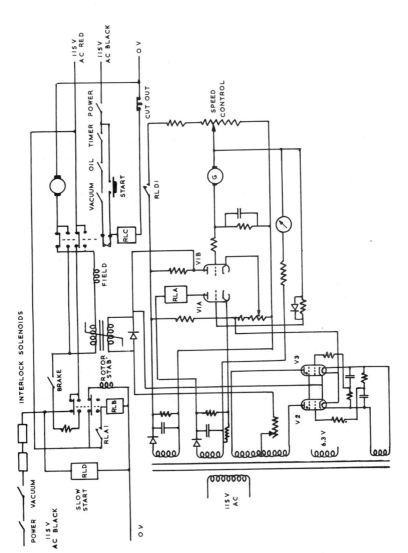

*Fig. 6.8.* Simplified circuit of Beckman Spinco Model L2 speed control.

applied to the motor. This may be done by using thyratron valves, a saturable reactor or silicon controlled rectifiers. The Beckman-Spinco model L ultracentrifuge used thyratrons and a saturable reactor to control the motor speed, but more modern instruments such as the Beckman L265B and MSE50 and 65 use solid state control circuits.

### Thyratron and saturable reactor motor speed control

A simplified circuit of the Beckman Spinco model L ultracentrifuge motor control system is shown in figure 6.8. The sequence of events after switching on and selecting a speed on the speed control potentiometer is as follows. A positive voltage, derived from the wiper of the speed control, is applied through the tachometer generator to the grid of the amplifier valve V1B. The valve conducts and a voltage drop is developed across its anode resistance, lowering the voltage applied to the cathodes of the thyratron valves V2 and V3. These valves which act as rectifiers then conduct and supply the d.c. control current for the saturable reactor. This d.c. current controls saturation of the saturable reactor core material and hence the impedance of the secondary winding. The current through the secondary winding increases and the motor accelerates. As the motor speed increases, the generator voltage output builds up until it is equal and opposite to the voltage obtained from the speed control wiper. The control system then maintains the speed at the selected value.

As the ultracentrifuge operating speed is high, the rotor runs in a refrigerated vacuum chamber and before the machine will run, vacuum and oil reservoir interlock switches must be closed. A safety device consisting of a knock out pin in the bottom of the the rotor chamber and a safety cartridge screwed into the base of the rotor chamber operates if the rotors maximum speed is exceeded. A plunger in the centre of the cartridge pops out and hits the knockout pin which opens the circuit to relay RLC. The model L2 has a solenoid operated rotor stabilising arm which damps the rotor during acceleration and deceleration so that rotor wobble is virtually eliminated. This device is useful for density gradient ultracentrifuge work.

While running the centrifuge up to speed the output from the generator 'bucks out' the negative bias on valve V1A and at approximately 1,200 r.p.m. drives its grid positive. This causes the valve to conduct and energise relay RLA, which in turn energises relay RLB lifting off the damper of the rotor stabiliser and priming the brake circuit. On completion of the centrifuge run, the preset timer returns to zero and relay RLC is de-energised. If the brake switch has been preselected, the a.c. supply to the drive motor is reversed and the motor dynamically braked in order to bring the rotor quickly to rest. When the speed has decreased to about 1,200 r.p.m. relays RLA and RLB are de-energised, as the positive output from the generator is less than the negative bias applied to valve V1A, which is then cut off. Relay RLB when de-energised actuates circuits which unlatch the chamber sliding door, de-activate the vacuum pump interlock and operate the damper of the rotor stabiliser.

**Transistor amplifier and magnetic amplifier motor speed control**

As an example for a solid state control circuit the MSE super speed 50 is described. The magnetic amplifier is of the auto excited type with rectifiers in series with the secondary load winding. The circuit is shown in figure 6.9, and all the safety interlocks (vacuum oil and lid) must be closed before relay RLA can be energised and voltage applied to the speed control potentiometer. With a speed selected on the potentiometer, a positive signal is applied to the base of transistor Q2 which conducts causing the collector voltages of the transistor pair Q1 and Q2 to be unbalanced. This gives rise to an input signal to transistors Q4 and Q5, which control the current in the d.c. winding of the magnetic amplifier. The impedance of the secondary windings is reduced and current flows through the motor circuit so that the rotor accelerates to the preset speed. If not limited, the motor current will rise to an excessive value causing damage to the motor itself, the magnetic amplifier windings and the excitation rectifiers. This is avoided by using a limiting circuit comprising resistance R1 and transistor Q3. The motor current flows through the low ohmic value resistance R1 and the voltage drop produced across it is applied to the base of transistor Q3. The emitter of transistor Q3 is biased so that the transistor conducts when the motor current approaches the desired limited value. The collectors of transistors Q1 and Q3 are effectively in parallel, the conduction of Q3 causing the control current in the magnetic amplifier to change so that the motor current is held at its limited value. This condition is maintained until transistor Q1 (the base of which is fed from the tachometer generator output) conducts and the motor current is reduced as the preset speed is approached. Transistor Q3 then stops conducting and the magnetic amplifier control current depends on the unbalance of the transistor pair Q1 Q2 and Q4 Q5. This in turn depends on the difference between the tachometer generator output and the voltage from the wiper of the speed control potentiometer. The motor speed is maintained constant at the preset value by the controlling action of the system. Should the speed tend to fall, transistor QI current is reduced and the current of transistor Q2 increased so that a signal is fed via transistors Q4 and Q5 to the magnetic amplifier control winding. The magnetic amplifier then alters the motor current so as to restore the speed to its original value.

The overspeed relay RLD is normally energised but should the machine overspeed, the centrifugally operated overspeed cut out switch on the motor shaft, operates and opens the relay circuit. A contact of relay RLD then opens the circuit of the motor control relay RLA and the centrifuge runs down.

Relay RLB primes the motor braking circuit and is controlled by the tachometer generator voltage and transistors Q6, Q7 and Q8. With no voltage output from the generator, transistor Q6 base is negative with respect to the base of transistor Q7. The collector current of Q6 is therefore very low and the transistor may be regarded as off. Current now flows through resistors R2 and R3 causing transistor Q8 to be turned on. This energises relay RLB but as the generator voltage rises a point is reached when transistor Q6 is turned on and Q7 off, this resulting in Q8 being turned off so de-energising RLB. At the end of the centrifuge run the manual/time switch is set to the off position, this causing relay RLA to be de-energised, and dynamic braking is then applied to the motor by reversing the armature current. When the speed falls to 1,000 r.p.m. relay RLB is energised and the braking current is removed. This allows the rotor to come to a stop slowly without disturbing the contents of the tubes in the rotor pockets.

## Motor control using silicon controlled rectifiers

The large capacity MSE Mistral 6L centrifuge uses silicon controlled rectifiers for the motor speed control, and a simplified circuit is shown as figure 6.10. The current through the motor circuit is controlled by the power silicon controlled rectifiers SCR2 and SCR3, these rectifiers being triggered by the pilot silicon controlled rectifiers SCR1 and SCR4. The current flowing through the motor is controlled by varying the firing point of the rectifiers in the positive half cycle of the supply voltage. This is performed by means of the phase shift network, comprising resistances R2, R3, R4, transformer T1 and the secondary winding LS of the magnetic amplifier. With full direct current in the magnetic amplifier control winding, the impedance of the secondary winding is low and the transformer T1 primary is effectively in parallel with R3. The waveform obtained when there is no d.c. control current (and the impedance of the magnetic amplifier secondary winding is high) is electrically out of phase with the previous case, so that by varying the impedance of the secondary winding of the magnetic amplifier we can control the firing point of the pilot silicon controlled rectifiers. The circuit is arranged so that when there is no d.c. control current, the silicon controlled rectifiers do not conduct.

The operation of common emitter transistor pair Q1 and Q2 is similar to that described for the MSE super speed 50 machine

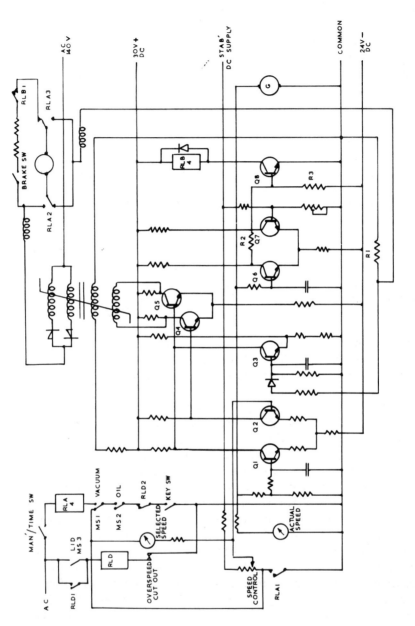

*Fig. 6.9.* MSE 50 ultracentrifuge speed control circuit.

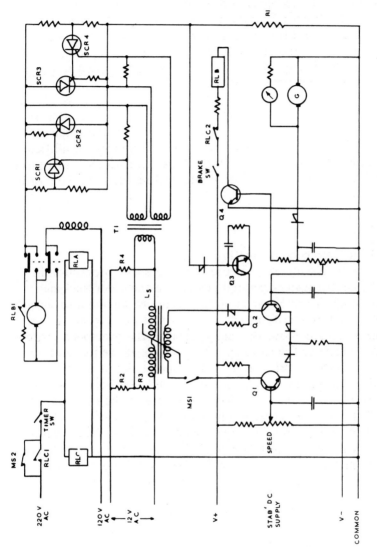

*Fig. 6.10.* Simplified speed control circuit of MSE Mistral 6L.

except that the control winding of the magnetic amplifier is directly connected to the transistor collectors. Transistor Q3 limits the motor current and transistor Q4 is conducting above a speed of 300 r.p.m. and energises relay RLB. This relay primes the braking circuit and is de-energised, removing the braking current, when the rotor speed falls below 300 r.p.m.

In the Beckman L2. 65B ultracentrifuge speed control circuit (figure 6.11), a toothed wheel, working in conjunction with a lamp and a photodiode, produces a train of pulses proportional to the drive speed. These pulses are amplified and passed through pulse shaping and a differentiating circuit in a tachometer amplifier assembly. The output from this amplifier is fed to the RPM indicating meter and to the speed control amplifier. In this unit, the motor speed signal is compared with a reference signal set by the RPM select potentiometer, and any difference is amplified and applied to a unijunction transistor circuit which controls the firing point of a pair of silicon controlled rectifiers in the motor drive circuit.

**Electronic overspeed protection**

Electronic circuits may be used for overspeed protection in place of centrifugally operated devices. In the older Beckman Spinco model L, a voltage approximately proportional to speed was obtained by rectifying the motor armature current. If this voltage increased above the bias applied to the overspeed protection valve (due to the rotor speed increasing above its maximum rated value) then a relay in the valves anode circuit operated and caused the centrifuge to run down. In the latest Beckman centrifuge the L265B, a recessed disc on the base of the rotor has alternate light and dark segments and works in conjunction with a photoelectric system to give overspeed protection. The output pulses from a photodiode are amplified and compared with a reference frequency (derived from a 31KHz crystal oscillator via a divide by two multivibrator circuit) in a comparator circuit. Should the rotor exceed its maximum speed by a margin of 2%, a relay is operated which causes the centrifuge to run down.

The Griffin Christ omega 11 has a unique speed limiting device which employs a toothed ring secured to the underside of the rotor. In the base of the rotor chamber there is a small coil with an iron core, and the pulses produced in the coil by the action of the rotating toothed ring are coupled to a circuit tuned to 12,300 Hz. Each rotor produces this frequency at approximately 500 r.p.m. above its maximum rated speed. If overspeeding occurs and the 12,300 Hz resonant frequency produced, a relay is operated which causes the centrifuge to run down.

## Speed control of analytical ultracentrifuges

In the analytical ultracentrifuge, a liquid sample is subjected to high controlled centrifugal force (up to approximately 372,200 'g') and the behaviour of solute molecules or suspended particles is recorded by photographic means using Schlieren, interference or absorption optics or by using a photoelectric scanner. In the Schlieren system, changes in refractive index gradients produced in the cell undergoing ultracentrifugation are measured. Solute concentrations between 0.1 and 2.0% are used, so that the change in refractive index is sufficient to give a large peak, which is caused by the boundary between the solute and sedimenting particles. Using the interference system, Raleigh fringe patterns are produced and more accurate results can be obtained. This system can be used with dilute solutions (0.05–0.5%). Ultraviolet optics are highly sensitive and concentrations as low as 0.001% can be measured.

Speed control methods used in these machines utilise different techniques to those employed on preparative laboratory centrifuges, as the rotor is usually suspended by wire or driven by an air turbine. The rotor runs in an evacuated and refrigerated chamber, provision being made for the several different optical systems to be used.

The original Beckman Spinco model E ultracentrifuge, used an electromechanical control system to regulate the speed of the wire suspended rotor to an accuracy of 60 r.p.m. in 60,000. This system uses a differential gear which continually compares the actual speed with the selected speed as referenced to a synchronous motor, the drive power being adjusted accordingly. An electronic control system has recently been introduced which gives 56 possible speed settings between 800 and 80,000 r.p.m. although the highest speed which at present can be utilised is 68,000 r.p.m. A tachometer generator, producing a voltage proportional to actual speed is used, and its output compared to a stable reference voltage. Any difference in the two voltages causes the drive power to change until the correct speed is reached, the average long term variations of rotor speed from the set figure being less than 0.1%

Electronic techniques may be applied to the air driven analytical ultracentrifuge, an example being the Hungarian Metrimpex model G120. This machine uses a frequency generator for speed measurement, the frequency being compared against tuned circuits to operate the control system which regulates the speed to 0.1% between 6,000 and 60,000 r.p.m.

The M.S.E. analytical ultracentrifuge uses a three phase motor which is driven from an electronic a.c. supply of variable frequency. The motor is supplied through three power amplifiers, and electronic discriminator circuits control the drive to an accuracy of one part in 10,000. The speed range is from 3,000 to 75,000 r.p.m. and both the motor and rotor are enclosed in the vacuum chamber, there being no gear box.

A speed control circuit for a magnetically suspended ultracentrifuge has been described by Beams (2). In this system, drive coils act as the field and the rotor neck as an armature. The speed of the rotor is measured by a photoelectric system, and the frequency of the pulses obtained is compared with the frequency divided output from an oscillator by utilising a difference amplifier. The amplifier output passes through diode and capacitor circuits to a mixer stage, the output of which feeds a phase splitter bridge. The bridge circuit output controls the drive coils.

### Ultracentrifuge photoelectric scanning systems

An automatic scanning system for absorption optics on the analytical ultracentrifuge has been described by Lamers et al. (9). A photomultiplier is slowly scanned across the image of the ultracentrifuge cell, which is of the double sector type. As the rotor rotates, a long dark period occurs followed by two light bursts, due to the solvent (reference) and the solution (sample) compartments in the cell. The first reference pulse is amplified and passed through a reference gate to a holding circuit. The trailing edge of this pulse closes the reference gate and opens a sample gate so that the following sample pulse is, after amplification, fed to a separate holding circuit. The amplifier section is common for both sample and reference pulses and includes a logarithmic circuit so that the signals held in the capacitor holding circuits are proportional to absorbance. The holding circuit outputs are fed to a difference amplifier, and the difference between the sample and reference pulses then recorded. A derivative amplifier also enables the derivative of optical density to be recorded.

A later system due to Lamers uses two double sector cells. Two photomultipliers are employed, one of which scans the pulsating images of the light from the monochromator which passes through the cell. The other photomultiplier is stationary, and gives a pulse output when light from the Schlieren source passes through the radius marker hole in the rotor and is deflected to the tube face by a mirror. A block diagram showing the principles involved in this and other circuits described is shown as figure 6.12. The stationary photomultiplier output pulse is used to control gating circuits in the measuring circuit sample and reference channels. The amplified pulses from the scanning photomultiplier pass through a logarithmic amplifier, and the sample and reference channel, to a holding circuit. This circuit feeds a differential amplifier the output of which is recorded (10).

# ELECTRONIC CONTROL OF ULTRACENTRIFUGES

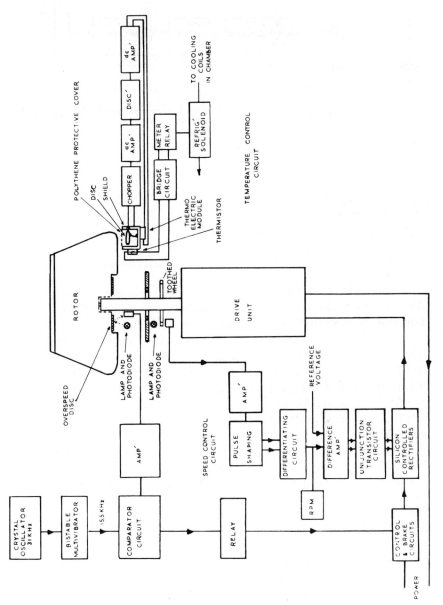

*Fig. 6.11.* Beckman L2.65B ultracentrifuge electronic control circuits.

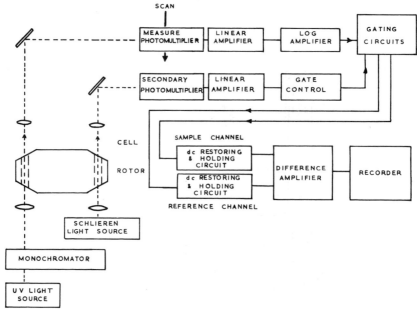

*Fig. 6.12.* Ultracentrifuge U.V. Scanner principle.

In the method of Spragg and Travers an oscillating mirror system is used and operates on the split beam principle (23). This system is similar to one designed by Schachman (20).

A split beam recording scanner for use with a multicell rotor and absorption optics has been described by Ping-Yao Cheng and Littlepage. The system can be used with 2, 4, or 6 cell rotors and utilises a multiplex system to select the solution and solvent to be scanned. This circuit also uses the Schlieren source and radius marker on the counter balance as a reference point (17).

Commercial ultracentrifuge U.V. scanners are available from M.S.E. Ltd. and Beckman Instruments.

Computer techniques can be applied to the ultracentrifuge data and will no doubt soon be incorporated into the ultracentrifuge. Trautman discusses this possibility and also centrifuge data processing programs (27).

### Temperature measurement and control

It is necessary to use precision electronic temperature measuring and control circuits on ultracentrifuges. Machines such as the Beckman Spinco model L2 and M.S.E. super speed 50, utilise a temperature measuring probe which is positioned close to but not in contact with the centre of the rotor. The Griffin Christ omega 11 overcomes the possibility of the probe being damaged by locating the temperature sensor inside the rotor drive shaft. This method ensures unimpeded thermal conductivity between the sensor and the rotor, inductive coupling between a pair of coils being used to make connection to the measuring circuit. One coil revolves with the rotor and the other outer coil is attached to the rotor chamber and forms one arm of a temperature measuring bridge circuit. The output of the bridge circuit is applied through an amplifying stage to the temperature indicator, causing the indicating needle to be deflected. The outside walls and base of the rotor chamber are wound with refrigeration coils which, when the machine is in operation, alternately heat and cool the evacuated chamber to control the preset temperature ($-2$ to $+25°C$) to an accuracy of $±0.2°C$.

The temperature indicator has in addition to the black temperature indicating needle, blue and red pointers which can be set as desired. The blue pointer is set to the required rotor temperature, and the red pointer is set to a higher value acting as a safety device should the temperature control fail. Both pointers have small selenium photocells attached, and temperature control is obtained by a mask on the indicating needle obscuring the light path between the blue pointer photocell and a small lamp. The photocell is connected to the base of a transistor which forms part of an electronic proportional controller. This controls the alternate heating and cooling of the rotor chamber so that the required temperature is reached exponentially.

The Beckman Spinco model L2 utilises two thermistors, one suspended in the hollow stem handle of the rotor by means of the stabilising arm; the other located next to the refrigeration liner in the vacuum chamber. They form part of an electronic bridge circuit which is regulated by a dial on the control panel and balanced for measurement by a second dial.

A temperature sensitive platinum resistor is used as the sensing element on the M.S.E. superspeed 40 and 50 model TC preparative ultracentrifuges, a simplified diagram of the electronic circuit being shown in figure 6.13. Variations of rotor temperature act on the temperature sensitive resistor RT and unbalance the bridge circuit, causing the indicating meter to deflect. The voltage difference between points A and B is chopped at mains frequency by transistors Q1 and Q2, which are turned on and off alternately by the centre tapped a.c. supply from transformer T1. This results in a square wave which is amplified by transistors Q3 and Q4. The output of the amplifier is applied to the trigger or gate electrode of the silicon controlled rectifier SCR1, which fires when the square wave input is in phase with the a.c. supply to relay RLA. Current flows through this relay on positive half cycles of the applied wave form, and relay RLA contacts actuate the refrigeration unit which controls the temperature to within $± 1°C$. Relay RLA is held in, on the negative half cycle of the applied voltage, by means of a capacitor.

252 ELECTRONICS IN ANALYTICAL SEPARATION TECHNIQUES

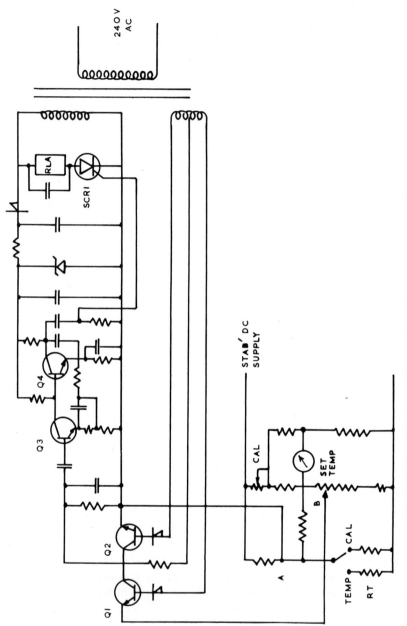

*Fig. 6.13.* Simplified circuit of MSE50 electronic temperature control.

Waugh and Yphantis have described a radiation thermocouple method which measures the infra-red radiation from a centrifuge rotor. Copper constantin thermocouples are utilised (29). An infra-red detection system is used on some of the latest ultracentrifuges and an example is the Beckman L2. 65B. (figure 6.11). This type of detector is very sensitive and has a fast response, so that good control of the temperature of samples in the rotor cavities is obtained.

A radiometer measures the rotor temperature, a shield defining the area from which infra-red radiation is received. The energy receiving element is an aluminium disc and if there is a temperature difference between the shield and the disc, an output signal is obtained from thermocouples attached to these two elements. This d.c. output is converted to a.c. by a chopping circuit, amplified, reconverted to d.c. and the amplified signal used to drive a thermoelectric module. This module is attached to the shield and either heats or cools it (depending on the direction of current flow) until it is at the same temperature as the disc. This temperature is then the same as the rotor temperature. A thermistor embedded in a block attached to the shield forms part of an electrical bridge circuit which feeds a meter-relay. There are two pointers on the meter and one is set to the required rotor temperature and the other to a temperature above which the rotor must

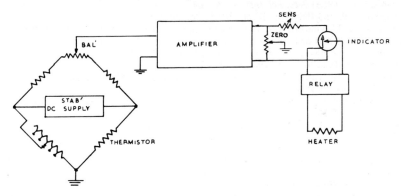

*Fig. 6.14.* Diagram of Beckman Spinco Model E temperature control.

not rise (safety circuit). When the indicating pointer rises above the required rotor temperature, contacts are closed and a solenoid is operated which permits refrigerant to flow and so cool the rotor. Temperature of samples in the rotor cavities are controlled to better than ± 1°C.

A miniature radio transmitter measurement and control system has been described by Fabricant, Windsor and Lindgreen (4). A transistorised Colpitts oscillator (FM) is inserted in a hole at the base of the rotor. The oscillator frequency increases with temperature at the rate of 60kHz per °C, due to the negative temperature coefficient of capacitors used in the oscillator tuned circuit. The unit is powered by mercury cells and a receiving loop antenna feeds a FM detector. The system measures to an accuracy ± 0.05°C.

Precise control of the temperature is also necessary in analytical ultracentrifuges. The Beckman Spinco model E uses a thermistor buried in the base of the rotor. A needle dips into a pool of mercury at the bottom of the rotor chamber to maintain contact between the thermistor and a bridge circuit, which in conjunction with built in refrigeration and heating units can control the temperature to an accuracy of ± 0.1°C. Any variation from the set temperature causes an unbalance in the bridge circuit which gives a signal to an amplifier unit. The output of the amplifier operates a relay through contacts on the indicator, the relay in turn actuating the heating element. This is shown in block diagram form in figure 6.14. The measurement and control of temperature, using thermistors on Beckman Spinco AN-D and AN-H analytical ultracentrifuge rotors, has been investigated by Gropper and Boyd (6).

### References

1. Analytical Chemistry (1953) **25**, 1423.
2. Beams, J. W. (1966) *Rev. Sci. Inst.* 37. **5**, 667.
3. Cropper, F. R. and Heinekey, D. M. (1967) Column No. 4., Vol 1. W. G. Pye Ltd.
4. Fabricant, S. J., Windsor, A. A. and Lingdreen, F. T. (1966) *Rev. Sci. Inst.* **37**, 495.
5. Gill, H. (1967) Analogue Dialogue. Vol 1. 2. 1. Analogue Devices Inc.
6. Gropper, L. and Boyd, W. (1965) *Anal. Biochem.* **11**, 2, 238.
7. Hughes, D. E. P. (1965) *J. Chem. Ed.* **42**, 450.
8. James, A. T., Rovenhill, J. R. and Scott, R. P. (1964) in Gas Chromatography (ed) A. Goldup. *Inst. of Petroleum.* 197.
9. Lamers, K., Putney, F., Steinburg, I. Z. and Schachman, A. K. (1963) *Arc. Biochem. and Biophysics.* **103**, 379.
10. Lamers, K. W. (1966) *Rev. Sci. Inst.* **37**, 11, 1616.
11. Lloyd, P. F. and Syers, K. (1964) *Lab. Prac.* **13**, 841.
12. Lovelock, J. E. (1958) *Chromatography.* **1**, 35.
13. Martin, A. J. P. and Jones, A. T. (1966) *Biochem. J.* **63**, 138.
14. Nerheim, A. G. (1963) *Anal. Chem.* **35**, 1640.
15. Noble, A. K. (1964) *Anal. Chem.* **36**, 8.
16. Pasuka, A. E. (1964) *Can. J. Biochem.* **42**, 219.
17. Ping Yao Cheng. and Littlepage, J. L. (1966) *Anal. Biochem.* **15**, 211.
18. Price, L. W. (1966) *Lab. Prac.* **15**, No. 3.
19. Robinson, D. W. (1965) *Dissertation abst.* **25**, 6921.
20. Schachman, H. K. (1963) *Biochemistry.* **2**, 887.
21. Shoemake, G. R., Fenimore, D. C. and Zlatkis, A. (1965) *J. Gas. Chromat.* **3**, 285.
22. Smith, L. R. (1967) Analogue Dialogue Vol 1. 2. 6. Analogue Devices Inc.
23. Spragg, S. P., Travers, S. and Santon, T. (1965) *Anal. Biochem.* **12**, 2, 259.
24. Stauffer, J. E., Kersten, T. E. and Krueger, P. M. (1964) *Biochem. and Biophys. Acta.* **93**, 191.
25. Stegemann, H. and Lerch, B. (1964) *Anal. Biochem.* **9**, 417.
26. Thorne, E. Z. (1964) *Instrumentenkunde.* **72**, 166.
27. Trautman, R. (1967) Fractions No. 2. Beckman Instruments.
28. Vesterberg, O. and Svensson, H. (1966) *Acta. Chemica. Scandinavica*, **20**, 820.
29. Waugh, D. and Yphantis, D. A. (1952) *Rev. Sci. Inst.* **23**, 609.
30. Yates D. (1967) Microtek Inc. at Techmation Ltd., Seminar, London.

# Illustration Index

For ease of reference by research workers, the instrument diagrams and circuits are listed under the appropriate analytical technique.

ATOMIC ABSORPTION

Unicam SP90 spectrophotometer, 122

ELECTRO-CHEMISTRY (GENERAL)

Beckman Electroscan 30, 74

ELECTRONIC DEVICES AND CIRCUITS

Amplifier,
  a.c., 2
  chopper, 13
  d.c., 12
  emitter coupled, 16
  feedback, 15
  operational, 19
  push pull, 11
Cold cathode stabiliser, 3
Diode,
  PN junction, 5
  tunnel, 5
  zener, 5
Klystron,
  principle, 23
  reflex, 23
Multivibrator, 187
Oscillator,
  Hartley, 21
  RC, 21
  wein bridge, 21
Rectifier,
  full wave, 24
  half wave, 24
  silicon controlled, 9
  voltage doubler, 24
Servo amplifier, 102
Stabilised d.c. supply, 25
Storage oscilloscope, tube, 132
Transistor,
  conventional, 6
  field effect, 6
  unijunction, 6

ELECTRON SPIN RESONANCE

Klystron automatic frequency control, 164
Lock-In amplifier E.S.R. spectrometer, 163
Pound stabiliser circuit, 165
Superheterodyne E.S.R. spectrometer, 163
Varian 4502 E.P.R. spectrometer, 166

EMISSION SPECTROMETRY

Hilger and Watts polyvac, 144

FLUORIMETRY

Aminco Bowman spectrophotofluorometer, 125

GAS CHROMATOGRAPHY

Beckman thermal conductivity detector, 224
Electron capture detector, 222
Flame ionisation detector, 222
Flame photometric detector, 222
Ionisation chamber gas chromatography, 178
LKB combined gas chromatograph–mass spectrometer, 152
Micro Tek ultrasonic detector, 226
Perkin Elmer parametric amplifier system, 229

LIQUID CHROMATOGRAPHY

Beckman 125 digital integrator, 236
LKB uvicord 11 column analyser, 232
Photo - Schmitt trigger fraction collector, 234

MASS SPECTROMETRY

LKB combined gas chromatograph–mass spectrometer, 152
Magnetic deflection mass spectrometer, 147

# ILLUSTRATION INDEX

## NUCLEAR MAGNETIC RESONANCE

Crossed coil N.M.R. spectrometer, 155
Feedback oscillator N.M.R. spectrometer, 156
High sensitivity field modulation N.M.R. spectrometer, 155
Perkin Elmer N.M.R. magnet environmental control circuit, 158
Perkin Elmer R10 N.M.R. spectrometer, 159
Superheterodyne detection N.M.R. Spectrometer, 157
Tuned R.F. bridge N.M.R. spectrometer, 156

## NUCLEAR PULSE SPECTROMETRY

Ansitron 11 scintillation counter, 208
Beckman CPM200 scintillation counter, 207
Beta pulse spectrum, 201, 205
Channels ratio quench correction, 211
Coincidence circuit, transistor, 196
Discriminator,
    Schmitt trigger, 204
    tunnel diode, 205
Nuclear Chicago
    anticoincidence circuit, 206
    6860 scintillation counter, 207
    tunnel diode coincidence circuit, 197
Nuclear Enterprises 8305 scintillation counter, 207

## NUCLEAR RADIATION DETECTION

Data print out, 190
Fast scaling circuit, 187
Gas flow counter, 177
Geiger counter, 176
Ionisation chamber gas chromatography, 178
Magnetic beam switching tube, 191
Panax
    AU460 anti coincidence unit, 192
    RM202 ratemeter, 186
    102ST scaler, 184
Ratemeter integrating circuit, 185
Solid state detector and ratemeter, 178

## pH MEASUREMENT

Beckman
    72 pH meter, 32
    research pH meter, 42
    zeromatic pH meter, 37
E. I. L. type 23A pH meter, 36
Leeds and Northrup 7402 pH meter, 38
pH–emf curve, 34
Pye
    290 pH meter, 38
    dynacap pH meter, 40
    Potentiometric pH meter, 44

## POLAROGRAPHY

Operational amplifier polarograph, 57
Oscillographic polarograph, 57
Polarographic waveforms, 66
Simple polarograph circuit, 57
Southern analytical Mk 11 pulse polarograph, 62

## POLARIMETRY

Bellingham and Stanley Polarmatic 62 spectropolarimeter, 129

## SPECTROPHOTOMETRY

Beckman
    DB spectrophotometer, 106
    DK2 spectrophotometer, 99, 104
    DU spectrophotometer, 92
Cary 15 spectrophotometer, 111
Hilger-Gilford spectrophotometer, 94 spectrochem, 93
Shimadzu AQV50 spectrophotometer, 96
Spectral response curves, light sources and detectors, 87
Unicam
    SP90 atomic absorption spectrophotometer, 122
    SP200 spectrophotometer, 99, 107
    SP500 spectrophotometer, 92
    SP500, series 2, spectrophotometer, 82
    SP800 spectrophotometer, 108
    SP3000 spectrophotometer, 112
Zeiss RPQ20A spectrophotometer, 110

## TITRATION

Coulometric titrator, 54
Pye Auto-titrator, 50
Radiometer titrigraph, 48
Sargent oscillometer, 71

## ULTRACENTRIFUGATION

Analytical ultracentrifuge U.V scanner, 250
Beckman
    L2 ultracentrifuge speed control, 240
    L265B ultracentrifuge speed and temperature control, 249
    Model E analytical ultracentrifuge temperature control, 253
M.S.E.
    Superspeed 50 ultracentrifuge speed control, 244
    Superspeed 50 ultracentrifuge temperature control, 252
    Mistral 6L speed control, 245

## X-RAY ANALYSIS

X-Ray fluorescence spectrometer, 170

# Subject Index

Absorption spectrophotometers, 81, 91
A.C. amplifier, 1
A.C. polarography, 60–67
Activation analysis, 218
A.E.I. MS10 mass spectrometer, 141, 153
Alkaline error (pH), 29
Amino acid analysers, 235
Aminco Bowman spectrophotofluorometer, 125, 126, 90
Amperometric titration, 52–53
Amperometric titrator circuits, 52
Amplification factor $\mu$, 1
Amplifier
  types of, 10
  basic a.c., 2
  cathode coupled, 16
  chopper, 14
  common base, 10
  common collector, 10
  common emitter, 10
  direct coupled, 12–14
  differential, 17, 18
  differentiating, 17
  electrometer, 14–15
  emitter coupled, 16
  feedback, 15–16
  integrating, 17
  lock in, 22
  logarithmic, 17
  operational, 17–18
  parametric, 227, 229
  paraphase, 17
  phase splitter, 17
  power, 11
  pulse, 202
  push pull, 11
  transformer coupled, 10
  tuned, 10
Analytical
  spectrometer systems, 141
  ultracentrifuges, 247, 248, 254
Analogue to digital conversion, 217
AND gate, 206
Ansitron 11 liquid scintillation counter, 208
Anti coincidence circuit, 193–195, 206
Antilogarithmic circuit, 30
Antimony electrodes, 29

Argon chromatograph, 221
Astable multivibrator, 201
Asymmetry potentials (pH), 31
Atomic absorption
  principle, 121
  spectrophotometers, 122
  resonance detector, 124
Autoburette unit, 49
Automatic
  standardisation pH meter, 35, 37
  sample changers, 209
Automation techniques, 75–76

Background
  flame, 120
  radiation detectors, 182–183
  liquid scintillation counting, 198–200
Baird Atomic Fluorispec, 126, 89, 90
Barretter, 33
Barrier layer photocell, 86
Baseline drift correction, 230
Bausch & Lomb precision spectrophotometer, 89
Beam switching tube, 191, 193
Beckman
  analytical ultracentrifuge, Model E, 247, 253
  digital integrator, Model 125, 235
  electroscan, Model 30, 74, 75
  fluorimeter, ratio, 126
  gas chromatograph, thermal conductivity detector, 224
  liquid scintillation counters, 183 207, 210
  mercury lamp, 85
  oxygen analyser, Model 777, 68
  oxygen electrodes, 68
  planchet gas flow counters, 183
  pH electrodes, 28
  pH meter,
    zeromatic, 35, 37
    research, 42, 43
    Model 72, 31, 33
  spectrophotometer,
    Model DB, 106, 109, 90
    Model DK2, 98, 99, 104, 105, 90
    Model DK, 117

# SUBJECT INDEX

Beckman—*continued*
  Spectrophotometer—*continued*
    Model DU, 92
    Model DU2, 116, 90
    Model IR4, 117
    Model IR5A, 116
    Model IR8, 91
    Model IR9, 91
    Model IR11, 116, 91
    Model IR12, 116, 91
    Model IR102, 133
  titrator, Model K, 51, 53
  ultracentrifuge,
    Model L2, 240, 241, 250
    Model L265B, 246, 249, 253
Beers law, 97
Bellingham and Stanley Polarmatic, 62, 128, 129, 90
Bennet radio frequency mass spectrometer, 151
Bistable multivibrator, 20
Blocking oscillator, 181
Bolometer, 85,
Bragg formula, 170
British standard 2586, 31

Calculators, electronic, 214
Canalco U.V. analyser, 231
Carbon dioxide electrode, 29, 30
Cary, spectrophotometer,
  Model 14, 90, 89
  Model 15, 111, 118, 117, 89
  spectropolarimeter, Model 60, 130, 90
  Raman spectropolarimeter, Model 81, 146, 141
Cathode ray tube, storage, 131, 132
Cathode ray polarographs, 64–67
Cation sensitive membranes, 29
Cationic electrodes, 29
Cells,
  spectrophotometric, 81
  electrochemical, 74
Channels ratio, quench correction, 212
Check point value (pH), 27
Chopper, vibrator, 39
Chopper amplifier, 14
Chloride ion electrodes, 29
Chromatography,
  gas, 221
  liquid, 230
  radio, 221
  start-stop, 153
Chronoamperometry, 73
Chronocoulometry, 76
Chronopotentiometry, 73
Circular dichroism, 130
Clark oxygen electrode, 67
Coincidence circuit,
  principle, 195
  transistor, 196
  tunnel diode, 197
Cold cathode tubes, 3

Column monitors, 231
Combined gas chromatograph–mass spectrometer, 151
Common
  base configuration, 10
  collector, 10
  emitter configuration, 10
  mode rejection, 17
Comparative polarography, 65
Compound external standard, 213
Computers, use with instrumentation, 75, 76, 144, 214
Computer for average transients (CAT), 136
Compton electrons, 213
Concentric electrodes, 28, 30
Condenser current (polarography), 56
Conductivity,
  measurement, 69–70
  instrument circuits, 69
  instrument performance, 76
  temperature compensation, 69
Consolidated Dynamics mass spectrometer, 141
Constant voltage transformer, 35
  current power supply, 26
  coulometry, 55
  potential coulometry, 53–54
Controlled potential polarography, 59–60
Corona stabiliser, 3
Coulometer, 53
Coulometric analysis, 53–55
Coulson electrolytic detector, 227
Counting circuits, 181–189
  geiger, 181
  gas flow, 175
  proportional, 176, 200
  scintillation, 179, 199
Current gain,
  alpha, 7
  beta, 7
Crystal,
  piezo electric, 21
  pockels effect, 127
  radiation detection, 215
Cycloid focusing mass spectrometer, 151

Damping, servo systems, 103
Darlington amplifier, 26
D.C.
  amplifiers, 12–14
  amplifier, pH meters, 31–35
Dekatron tube, 181
Demodulation, 22
Derivative
  polarography, 59
  titrators, 51, 52
Detectors,
  electronic, 22
  for E.S.R., 162

## SUBJECT INDEX

Detectors—*continued*
   gas chromatography, 221–227, 228
   infra red, 84
   mass spectrometer, 149
   for N.M.R., 154
   radioactivity, 175–181, 215
   in spectrophotometry, 86–89
   x-ray, 170
Differential amplifier, 17, 18
Differentiating circuits, 17
Diffusion current (polarography), 56, 64
Digital
   integrator, 235
   pH meter, 45
   spectrophotometers, 95, 112, 114, 115
   techniques in mass spectrometry, 153
Diode,
   rectifier, 2, 4
   semiconductor, 4
   thermionic, 1
   tunnel, 5
   zener, 4
Discharge detector (gas chromatography), 228
Discriminators,
   adjustable amplifier bias, 203
   Schmitt trigger, 203
   tunnel diode, 204
Doran conductance meter, 76
Double focusing mass spectrometer, 150
Drift,
   amplifier, 13
   klystron, 164
   temperature, 13
Drop counters, 233
Dropping Mercury Electrode (DME), 52, 56
Durrum Instrument Co. stopped flow spectrophotometer, 90, 135

Ekco (E. K. Cole)
   N664 scintillation counter, 198
   N701 ratemeter, 223
Electroanalytical techniques, 72–76
Electrochemical instruments,
   multipurpose, 73–76
   performance, 76–77
Electron capture detector, 179, 222, 223
Electrodes,
   amperometric titration, 52
   carbon dioxide, 29
   characteristics (pH), 28
   combined (pH), 28, 30
   correct use of, 30–31
   dropping mercury, DME, 52, 56
   for emission spectrography, 142
   glass (pH), 28
   hanging drop mercury, H.D.M.E., 59
   metal, 29
   oxygen, 67
   pH measurement, 28
   reference, 30

Electrodes—*continued*
   specific ion, 29
   for titrations, 47, 52
Electrofocusing columns, 239
Electrolysis, 72
Electrometer amplifiers, 14, 15, 149
Electronic Instruments Ltd. (E.I.L.),
   Model 15A oxygen analyser, 77
   Model 23A pH meter, 33–35, 36, 46
   Model U28 ultrasonic electrode cleaner, 30
   Model 38A pH meter, 46
   Model pH electrodes, 28
   Model Vibret pH meter, 41
   Model Vibron pH meter, 41, 46
Electron
   multiplier, 149
   paramagnetic resonance spectroscopy, 154 (see electron spin resonance)
   spin resonance (E.S.R.), 154
E.S.R. spectrometers,
   a.c. modulation, 161
   heterodyne, 162
   lock in amplifier, 163
   microwave circulator, 168
Electrophoresis,
   continuous flow, 239
   materials, 238
   moving boundary, 239
   power supplies, 238
   principle, 237
Electrostatic effects (pH measurement), 31
Emission
   spectra, 87, 85
   spectroscopy, 139
   spectrographs, automatic, 144
      characteristics of, 140
      direct reading, 143
      photographic, 143
Eppendorf photometer, 98, 90, 124
Excitation in spectroscopy, 139
External standard quench correction, 212

Faradic current (Polarography), 56
Faraday cell, 128
Feedback,
   15, 16
   ratio $\beta$, 16
Field effect
   transistor, 8
   transistor pH meter, 43–45
Filters,
   electronic, 25
   photometric, 90, 119
Figure of Merit, 209
Flame
   ionisation detector, 222, 223
   photometry, 119
   photometers, integrating, 120
   photomultiplier, 119
   separate light path, 120

Flame—*continued*
   photometric detector (gas chromatography), 227
Flashing light spectrophotometer, 135
Flicker effect, 19
Flip flop, 20
Fluorimeters, 124
Fluorimeters,
   filter, 124
   double monochromator, 126
   phase and polarisation, 127
   ratio recording, 127
Four layer semiconductor devices, 9
Fraction collectors, 233
Frequency changing, 22
Full wave rectifier, 24

Gamma
   radiation detectors, 215
   spectrometers, 216
Gas
   chromatography, detectors, 221–227, 228
      dual column, 230
      principle, 221
      sub ambient, 229
      systems, 227
   density detector, 225
   filled valves, 2
   flow counting, 175
   plasma, 142
Geiger
   tube, 175
   counter
      characteristics, 182
      circuit, 181–185
Gilford spectrophotometer, 94
Glass electrodes, 28
Globar, 84
Golay detector, 85
Gold electrodes, 29
Grid current, 14
Griffin Christ omega 11 ultracentrifuge, 246

Half wave rectifier, 24
Hanging drop mercury electrode, HDME, 59
Harmonic polarography, 61
Heptode valve, 22
Hexode valve, 22
High frequency
   techniques, 21
   titrators, 70–72
Hilger Gilford spectrophotometer, 93, 89
Hilger & Watts,
   infrascan spectrophotometer, 117, 91
   polyvac spectrograph, 140, 144
   spectrochem, 90, 89, 93
   spekker, 90
   ultrascan, 89
   x-ray spectrometer, 140

Hitachi—Perkin Elmer,
   mass spectrometer, Model RMU–6E, 141
   spectrophotometer, Model 139, 97
Hollow cathode lamps, 121

Infra-red
   sources and detectors, 84
   spectrophotometers, characteristics, 91
   rapid scan techniques, 132
Infotronics corp, gas chromatograph data processing, 230
Instrumentation laboratories flame photometer, Model 143, 120
Insulators, 151
Interelectrode capacitance, 2
Interelement correction, 145
Integrated circuits, 9–10
Integrating
   amplifiers, 17, 53, 230
   motors, 53
Intermediate frequency I.F., 22
Internal
   standardisation, photometry, 119
   scintillation counting, 212
Intertechnique liquid scintillation counter, 183
Interrupted elution chromatography, 153
Intrinsic stand off ratio (n), 8
Ion source, 148
Ionisation
   chambers, 177
   detectors (gas chromatography), 178
Isco U.V. analyser, 223
Isotope ratio mass spectrometer, 150

Klein Eisler Number, 209
Klystron,
   frequency stabilisation, 164, 165
   principle, 22
   reflex, 23
   use in N.M.R., 162

Lamberts law, 97
Lamps,
   deuterium, 81, 83
   hollow cathode, 121
   hydrogen, 81, 83
   mercury, 84, 85
   quartz-iodine, 83
   torronto, 146
   tungsten, 81
   xenon, 84
Lamp power supplies, 83
Lamp spectral characteristics, 81
Lasers, 136, 142, 146
Leeds & Northrup pH meters, 38, 46
Light sources, spectrophotometric, 90, 91 81, 83, 84, 85, 121
Limiting current (polarography), 56
Lingane titrator, 50

# SUBJECT INDEX

LKB
  conductolyser, 76
  electro focusing column, 239
  gas chromatograph-mass spectrometer 9000, 141, 149, 151, 152
  uvicord, 231, 232
Lock in amplifier, 22
Logarithmic amplifiers, 17
Long-tailed pair, 17
Liquid chromatography,
  instrumentation, 231–237
  principle, 230
Liquid scintillation counting,
  principle, 179
  quenching, 211
  quench correction, 212
  single photomultiplier systems, 208
  two photomultiplier systems, 209–211
Live timing, 211

Mackerith oxygen probe, 68
Magnification factor, Q, 11
Magnetic amplifier, use in speed control, 242
Magnet environment control, 159, 160
Magnet superconducting, 157
Magnetic deflection mass spectrometer, 148, 150
Mass spectrometry, 146–153
Mass spectrometer,
  amplifiers, 149
  inlet systems, 147
  ion sources and collection, 148
  recording systems, 149
  regulating circuits, 149
  types of, 150
Measuring and Scientific Equipment Ltd., MSE,
  analytical ultracentrifuge, 247
  mistral 6L centrifuge, 243, 245
  superspeed 50 ultracentrifuge, 242, 244, 250
Microcoulometer, 266
Microminiature techniques, 10
Micro oxygen electrodes, 68
Microwave spectroscopy, 161 (see Electron Spin Resonance)
Microwave spectrometers,
  a.c. modulation, 162
  heterodyne, 162
  lock in amplifier, 163
  microwave circulator, 168
Microwave techniques, 22–24
Microtek Instruments Inc,
  sub-ambient gas chromatograph, 229
  ultrasonic detector, 225
Migration current (polarography), 56
Monochromators,
  double, 97, 115
  multipass, 97
  series, 97
Monostable Multivibrator, 20

Mössbauer spectrometry, 218
Mullard geiger tubes, 182
Multipot system, 118
Multichannel analysers, 217
Multichannel analysis, 204
Multivibrators, 20

Negative feedback, 15
Nernst glower, 84
Neutron generator, 218
Nier ion source, 148
Nichrome infra red source, 84
Noise,
  sources, 19
  effect on microwave systems, 164
Nuclear counters, 175–219
Nuclear Chicago,
  anticoincidence circuit, 206
  gas flow detector D47, 182
  liquid scintillation counter,
    Model 6725, 183
    Model 6850, 183
    Model 6860 (Mark 1) 207, 183
  low background spectroshield planchet counter, 182
  tunnel diode coincidence circuit, 196, 197
Nuclear Enterprises,
  gamma detector Model 5502, 182
  liquid scintillation counter,
    Model 8305, 207
    Model 8306, 183
  sample changer scintillation counter, Model 8310, 208
  spectrometer (integrated circuit), Model 8622, 218
Numerical indicator tubes, 188, 189
Nuclear Magnetic Resonance N.M.R., 154–161
N.M.R. spectrometers,
  crossed coil, 155
  feedback oscillator, 156
  field modulation, 155
  magnet control, 157
  pulsed, 161
  repetitive scan, 161
  tuned bridge, 156

Oak Ridge National Laboratory Titrators, 48–49
Olivetti Programma 101 desk-top computer, 209
Omegatron mass spectrometer, 153
Operational amplifiers,
  principles, 17, 18
  use in controlled potential electrolysis, 54
  use in polarography, 57, 59, 60
Optical null spectrophotometers, 99, 109
Optica spectrophotometers, Model CF4, densitronic, 90, 116
OR gate, 193

# SUBJECT INDEX

Oscillators,
  blocking, 181
  colpitts, 20
  crystal, 21
  Hartley, 20
  relaxation, 20
  sinusoidal, 20
  wein bridge, 20
Oscilloscope, storage, 131, 132
Overspeed protection, in ultracentrifuges, 246
Oxygen analysers, 68, 77
Oxygen electrode,
  operating techniques, 67–68
  principle, 67

Packard Instruments,
  liquid scintillation counters,
    3000 series, 183
    4000 series, 183
  compound external standard, 213
Panax Ltd.
  anticoincidence unit AU460, 192, 194
  ratemeter RM202, 185, 186
  scaler 102ST, 181, 184
  x-ray analyser, 172
Parametric amplifiers, 227, 229
Pentode valve, 2
Perkin Elmer Ltd,
  fluorescence spectrophotometer, 127
  gas chromatograph amplifier, 227
  polarimeter, Model 141, 90
  Raman spectrophotometer, LR-1, 141
  mass spectrometer, Model 270, 141
  N.M.R. spectrometer R10, 158–160 141
  N.M.R. spectrometer R12, 141
pH electrodes, 27–31
pH-emf equation, 27
pH-measurement, 27–46
pH meters,
  d.c. amplifier, 31–35
  digital, 45
  field effect transistor, 43–45
  performance of, 46
  photo conductive modulator, 38–39
  potentiometric, 43
  vibrating capacitor, 41–43
  vibrating chopper, 35–39
pH-Stat, 49, 50
Phase angle polarography, 61
Phase selective polarography, 61
Phase sensitive detector, 13, 14
Phase splitter, 17
Philips,
  conductance meter, Model 9500, 76
  gas chromatograph, Model 4000, 153
  pH meters,
    Model 9401, 46
    Model 9408, 45
  scintillation counter, 183
  titrator, Model 9450, 51, 52

Philips—*continued*
  x-ray analysers, 140
  x-ray analyser, Model 1250, 171
Phoenix
  dual wavelength spectrophotometer, 115
  light scattering photometer, 89
Phosphors, 179
Photocathodes,
  types, 86, 180
  S response, 88, 89, 180
Photocells,
  barrier layer, 86
  photoconductive, 88
  photoemissive, 86
  spectral response, 87
Photoconductive cell, 88
Photoconductive modulator, 39
Photoelectrons, 88, 173
Photoelectron spectrometer, 172
Photometers,
  double beam, 86
  filter, 86
  flame, 119
Photomultiplier,
  effect of cooling, 195
  matching with liquid scintillation solute, 180
  principle, 88
  use in flame spectrophotometers, 119
  use in liquid scintillation counting, 180, 195
  use in U.V. spectrophotometers, 89
Picker Nuclear DIRAC system, 215
Piezo electric crystal, 21
Platinum electrodes, 29, 69
PNPN semiconductor devices, 9
Pockels effect crystal, 127
Polarimetry, 127
Polarimeters, 127
Polarmatic 62 spectropolarimeter, 128, 129
Polarogram waveforms, 66
Polarograph circuits, 56–57
Polarographic oxygen probes, 67–68
Polarography,
  a.c., 60–67
  cathode ray, 64–47
  comparative, 65
  controlled potential, 59–60
  conventional, 55–56
  derivative, 59
  harmonic, 62
  operational difficulties, 58
  phase angle, 61
  phase selective, 61
  pulse, 62–64
  subtractive, 65
  square wave, 63
Polychromator, 143
Positive feedback, 19
Potentiometric pH meters, 43
Potentiometric titration, 47–52

# SUBJECT INDEX

Potentiostat, 72
Pound stabiliser, 165
Power amplifiers, 11
Power supply circuits, 24–26
Power units, electrophoresis, 238
Pre-amplifiers,
 use in spectrometry, 149
 use in spectrophotometry, 101
 use in scintillation counter, 200
Primary coulometry, 53
Print-out circuit, 189–193
Proportional control, 49
Proportional radiation counter, 200
Pulse amplifiers, 202
Pulse height analysis, 203–206
Pulse spectrum,
 beta radiation, 201, 205
 gamma radiation, 200
Pulse summation, 209
Push-Pull output, 11
Pye-Ingold pH electrodes, 28
Pye W. G. Ltd,
 conductance meter, Model 11700, 76
 liquid chromatography system, 235
 pH meters,
  Model 79, 33
  Model 290, 38, 45
   11607, 46
   Dynacap, 40, 41, 46
   potentiometric, 43, 44, 46
 argon radiochromatograph, 221
 Titrator, Model 11602, 50, 51, 76

Quantum efficiency, 195
Quartz crystal oscillator, 21
Quench correction methods, 212
Quenching gas, 175
Quenching liquid scintillation counting, 211

Radelkis OK302 oscillatrator, 70, 76
Radiation detectors,
 gas flow, 175, 177
 geiger, 175, 176
 ionisation chamber, 177, 178
 liquid scintillation, 179
 solid scintillator, 179
 solid state, 181
Radiochromatography, 221
Radiochromatograph, argon, 221
Radio frequency spectroscopy, 154
 (see Nuclear Magnetic Resonance)
Radio frequency spectrometers,
 crossed coil, 155
 feedback oscillator, 156
 field modulation, 155
 magnet control, 157
 pulsed, 161
 repetitive scan, 161
 tuned bridge, 156

Radiometer,
 conductance meter CDM2, 76
 pH electrodes, 28
 pH meter, Model PHM28, 46
 polariter, Model PO4, 77
 titrator TTT1, 48, 49, 76
 titrigraph, 48, 49, 76
Raman spectrometry, 145–146
Rapid scan techniques, 131
Ratemeter,
 circuit, 186
 principle, 185
Ratio recording spectrophotometer, 98
Recorder, servo amplifier, 103
Recording titrators, 47–50
Rectifier circuits, 24
Reference electrodes, 28, 30
Reflex klystron, 23, 24
Residual current (polarography), 55
Resonance detectors (atomic absorption), 124
Resonant circuit, 10, 11

Sargent
 oscillometer, 70–72, 76
 polarograph Model XV, 57, 77
 recording titrator, 76
Saturable reactor, 241
Scale expansion, 41
Scalers, 188
Scaler display, 188
Scaling circuits, 188, 187
Schmitt trigger circuit, 20, 160, 233
Scintillation counters, 183, 206–211, 216–218
Scintillation counting, 179
Scintillators
 solid, 215
 liquid, 179
Second derivative titrators, 52
Second derivative polarographs, 59, 65
Secondary coulometry, 53
Secondary electrons, 2
Semiconductors, 4–9
Severinghauss electrode, 29–30
Shielding, 15
Shimadzu spectrophotometer, Model AQV-50, 95, 96, 117
Shot noise, 19
Signal enhancement, 136
Silicon controlled rectifier,
 principle, 9
 speed control circuits, 243
Sinusoidal oscillators, 20
Smoothing, 25
Sodium electrode, 29
Sodium sulphate electrode, 30
Solid state radiation detectors, 181
Solutes, 180
Southern Analytical Ltd.
 cathode ray Polarograph,
  Model A1660, 77

## SUBJECT INDEX

Southern Analytical Ltd.—*continued*
  cathode ray Polarograph—*continued*
    Model A1670, 65
    flame photometer, Model 1740, 65
Southern–Harwell, pulse polarograph A1700, 64, 77
Specific ion Electrodes, 29
Spectrographs,
  automatic, computerised, 144
  characteristics, 140
  direct reading, 143
  photographic, 143
Spectrometers,
  beta radiation, 183, 206–211
  emission, 139–145
  E.S.R., 161–168
  gamma radiation, 216–218
  mass, 146–153
  microwave, 161–168
  Mössbauer, 218, 219
  N.M.R., 154–161
  photo electron, 172, 173
  radio frequency, 154–161
  Raman, 146
  x-ray, 169–172
Spectrometry, 139–174
Spectrophotometers,
  absorption, 91
  atomic absorption, 121
  digital, 95
  direct reading, 92
  double beam, 98
  double monochromator, 115
  flame, 119
  flashing light, 135
  infra-red, 91
  modifications to, 94, 115
  null balance, 91
  optical null, 100
  performance characteristics of, 116, 117
  rapid scan, 131
  ratio recording, 98
  stopped flow, 133
  system performance of, 101
  ultraviolet-visible, 90
Stabilised power supplies, 25
Stallwood jet, 139
Stark effect, 162
Stopped flow spectrophotometer, 133
Storage oscilloscope, 131, 132
Stray light, 97
Sub ambient chromatography, 229
Subtractive polarography, 65
Superconducting magnets, 157
Superheterodyne detection, 157, 163, 164
Synchros or selsyns, 48
Synchronous chopper, 39

Techtron atomic absorption spectrophotometer,
  Model, AA4, 90
  Model AR200, 124

Temperature compensation,
  conductance measurement, 69
  pH measurement, 33–34
Temperature control circuits, 250–254
Temperature stabilisation,
  amplifiers, 7
  N.M.R. magnets, 159, 160
Tektronix storage oscilloscope, 132
Tetrode valve, 2
Thermal conductivity detector, 224
Thermal noise, 19
Thermionic valves, 1
Thin film circuits, 9
Thyratron, 3
Thyratron control circuits, 49, 51
Thyratron speed control circuits, 241
Thyristor, 9
Time of flight mass spectrometer, 151
Titration,
  amperometric, 52–53
  potentiometric, 47–52
Titrators,
  burette closure, 50
  end-point derivative, 51–52
  high frequency, 70–72
  recording, 47–50
Torronto lamp, 146
Touzart et Matignon high frequency titrimeter, 76
Tracerlab scintillation counter, 210
Transistors,
  field effect, 8
  metal oxide silicon, 6, 9
  N.P.N., 6, 7
  P.N.P., 7
  thin film, 9
  Unijunction, 8
Trigger tubes, 4
Triode valve, 1
Trochotron tube, 191, 193
Truth table, 189
Tunnel diode,
  coincidence circuit, 196, 197
  discriminator, 204, 205
  principle, 5

Ultracentrifuge,
  analytical, 247
  preparative, 240
  photoelectric scanner for, 248
  speed control circuits, 239–248
  temperature control of, 250–254
Ultrasonic electrode cleaning, 30
Ultrasonic detector (gas chromatography), 225
Ultraviolet chromatograph column monitors, 231–233
Ultraviolet lamps, 83–85
Ultraviolet-visible spectrophotometers, 90
Unicam Ltd., spectrophotometer,
  Model SP90, 122, 90, 89
  Model SP200, 99, 100, 107, 109, 116, 91
  Model SP500, 92

Unicam Ltd., spectrophotometer—
*continued*
   Model SP500 series 2, 82, 116, 90
   Model SP600, 116, 90
   Model SP700, 117, 90, 89
   Model SP800, 108, 116
   Model SP900, 89
   Model SP1200, 91
   Model SP3000, 112, 114, 117, 90, 89
Unijunction transistor, 8

Varactor diode, 227
Varian
   E.P.R. spectrometer, Model 4502, 166–167
   E.S.R. spectrometer, Model E3, 168, 141
   digital integrator, Model 475, 230
   N.M.R. spectrometer, Model A-60, 141
     HA-100, 141
     4311, 157
Vibrating capacitor, 41
Vibrator, 39
Voltammetry, 74, 75
Voltage doubler, 24, 25
Voltage multiplier, 25

Warner & Swasey, rapid scanning spectrometer, 89
Waveguides, 22

Wavelength, spectrophotometer, 90, 91, 116, 117
Wavelength shifter, 179
Wein bridge oscillator, 20, 21
Window,
   radiation detector, 176
   pulse height analyser, 216, 212, 213

Xenon lamps, 84
Xenon lamp power supply, 125, 126
x-ray
   absorptiometer, 172
   automatic analysis, 171
   detectors, 170
   tubes, 168
   tube current control, 169
   spectrometers, 169

Yellow springs,
   oxygen analysers,
     Model 52, 77
     Model 53, 77
   oxygen meter, 68

Zeiss, Carl, spectrophotometer
   Model DMR21, 90
   Model RPQ20A, 110, 113
Zener diodes, 4, 5
Zero emf point (pH), 27, 28

*Printed Offset Litho in Great Britain by*
*Spottiswoode, Ballantyne and Co. Ltd.*
*London and Colchester*